Shape Memory Alloys

edited by
Corneliu Cismasiu

Shape Memory Alloys

Edited by Corneliu Cismasiu

Republished by InTech

Janeza Trdine 9, 51000 Rijeka, Croatia

Publishing Process Manager Iva Lipovic

Technical Editor Goran Bajac

Cover Designer Martina Sirotic

Additional hard copies can be obtained from orders@intechopen.com

Shape Memory Alloys, Edited by Corneliu Cismasiu
p. cm.
ISBN 978-953-307-106-0

For EU product safety concerns:
IN TECH d.o.o., Prolaz Marije Krucifikse Kozulić 3, 51000 Rijeka, Croatia,
info@intechopen.com or visit our website at intechopen.com.

Contents

Preface

The Shape Memory Alloys (SMAs) represent a unique material class exhibiting peculiar properties like the shape memory effect, the superelasticity associated with damping capabilities, high corrosion and extraordinary fatigue resistance. Due to their potential use in an expanding variety of technological applications, an increasing interest in the study of the SMAs has been recorded in the research community during the previous decades.

This book includes fundamental issues related to the SMAs thermo-mechanical properties, constitutive models and numerical simulation, medical and civil engineering applications and aspects related to the processing of these materials, and aims to provide readers with the following:

- It presents an incremental form of a constitutive model for shape memory alloys. When compared to experimental tests, it proves to perform well, especially when the stress drops during tension processes.
- It describes single-crystal and multi-grained molecular models that are used in the dynamic simulation of the shape memory behaviour.
- It explains and characterizes the temperature memory effect in TiNi and CU-based alloys including wires, slabs and films by electronic resistance, elongation and DSC methods.
- It analyses the thermo-mechanical behaviour of superelastic NiTi wires from low to impact strain rates, including the evolution of the phase transformation fronts.
- It presents an experimental testing programme aimed to characterize the bending deformation and the fatigue properties of precision-casted TiNi SMA used for instruments in surgery operations.
- It introduces a computational model based on the theory of hysteresis operator, able to accurately characterize the non-linear behaviour of SMA actuators and well suited for real-time control applications.
- It describes the development and testing of an SMA-based, highly sophisticated, lightweight prosthetic hand used for multifunctional upper-limb restoration.
- It presents the design of an active catheter and endoscope based on shape memory alloy actuators, expected to allow low cost endoscopic procedures.
- It exemplifies the use of superelastic shape memory alloys for the seismic vibration control of civil engineering structures, considering both passive and semi-active devices.
- It discusses some aspects related to the processing of the shape memory alloys and presents special techniques that provide bonds without severe loss of the initial SMA properties.

With its distinguished team of international contributors, Shape Memory Alloys is an essential reference for students, materials scientists and engineers interested in understanding the complex behaviour exhibited by the SMAs, their properties and potential for industrial applications.

Lisbon, July 2010

Editor

Corneliu Cismasiu
Centro de Investigação em Estruturas e Construção - UNIC
Faculdade de Ciências e Tecnologia
Universidade Nova de Lisboa
2829-516 Caparica
Portugal

1

Molecular Dynamics Simulation of Shape-memory Behavior

Takuya Uehara
Yamagata Univ.
Japan

1. Introduction

Mechanical properties of shape-memory alloys (SMAs) are typically represented by the characteristic stress–strain curve, which forms a hysteresis loop in a loading, unloading and shape-recovering process. To represent the deformation behavior of SMAs, various constitutive equations have been developed, and prediction of the macroscopic behavior has been possible using finite-element simulations. The atomistic behavior leading to the deformation and shape-recovery is explained on the basis of the phase transformation between austenite and martensite phases and the characteristics of the crystal structure.

One well-known atomistic mechanism is illustrated in Fig. 1. The stable phase depends on the temperature, and phases at high and low temperature are body-centered cubic (bcc or B2) and martensite, respectively. The martensite phase consists of many *variants*, and each variant has a directional unit cell. In Fig. 1(b), for example, a unit cell of the martensite is illustrated as a box leaning in the positive or negative direction along the x-axis. Cells leaning in the same direction constitute a layer, and the direction of the lean alternates between layers. In this paper, the layer is called a variant, although a realistic variant is defined as a rather larger domain. The martensite phase is generated by cooling the B2 structure shown in Fig. 1(a). Randomly orientated variants are then generated, as shown in Fig. 1(b). When a shear load is imposed on this state, some of the layers change their orientation, as shown in Fig. 1(c). This structural change induces macroscopic deformation. When the external shear load is released, the strain does not return to the original state except for slight elastic recovery. When the specimen is heated to the transformation temperature, the martensite transforms into the B2 structure, and martensite appears again with cooling of the specimen. Since the B2 structure is cubic, the shape of the unit cell is independent of the orientation of the martensite layers. Therefore, the specimen macroscopically regains its original shape.

This mechanism is well known but has not been fully verified since direct observation of dynamic behavior in a wide range of temperatures is difficult. Therefore, computer simulation is expected to provide evidence for and further extend the mechanism. The molecular dynamics method has become a powerful and effective tool to investigate material properties and dynamic behavior on an atomistic scale, and it has also been applied in the case of SMAs. The stable structure of Ni_3Al, for instance, was investigated by Foiles and Daw (Foiles & Daw, 1987), Chen et al. (Chen et al., 1989) using an interatomic potential based on the embedded-atom method (EAM) with suitable parameters (Daw & Baskes, 1984; Foiles et al., 1986). The phase stability and transformation between B2 and martensite structures in NiAl was also

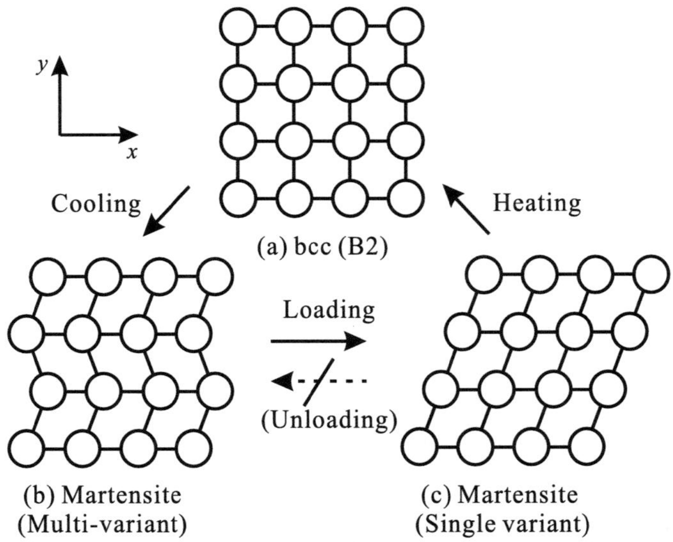

Fig. 1. Schematic illustration of deformation and shape recovery of a SMA.

reproduced using the EAM potential as reported by Rubini and Ballone (Rubini & Ballone, 1993) and Farkas et al. (Farkas et al., 1995). Uehara et al. then utilized the EAM potential to demonstrate the shape-memory behavior of Ni-Al alloy in terms of a small single crystal (Uehara et al., 2001; Uehara & Tamai, 2004, 2005, 2006), the size dependency (Uehara et al., 2006), and the polycrystalline model (Uehara et al., 2008, 2009). Ozgen and Adiguzel also investigated the shape-memory behavior of Ni-Al alloy using a Lennard-Jones (LJ) potential (Ozgen & Adiguzel, 2003, 2004). In addition, for Ni-Ti alloy, martensitic transformation was simulated by Sato et al. (Sato et al., 2004) and Ackland et al. (Ackland et al., 2008). It was also reported by Kastner (Kastner, 2003, 2006) that the shape-memory effect can be represented even by a two-dimensional model with a general LJ potential on the basis of thermodynamical discussion on the effect of temperature on the phase transformation. For a more practical purpose, Park et al. demonstrated shape-memory and pseudoelastic behavior during uniaxial loading of an fcc silver nanowire, and discussed the effect of the initial defects and mechanism of twin-boundary propagation (Park et al., 2005, 2006).

In this chapter, atomistic behavior and a stress–strain diagram obtained by molecular dynamics simulation are presented, following our simulation of Ni-Al alloy. A summary of the molecular dynamics method as well as EAM potential is given in Sec. 2. The simulation conditions are explained in Sec. 3. Simulation results obtained using the single-crystal model and polycrystal model are presented in Sec. 4 and 5, respectively, and concluding remarks are given in Sec. 6.

2. Molecular Dynamics Method

2.1 Fundamental equations

Employing the molecular dynamics (MD) method, the position and velocity of all atoms considered are traced by numerically solving Newton's equation of motion. Various physical and mechanical properties as well as dynamic behavior on the atomistic or crystal-structure scale are then obtained using a statistical procedure.

The fundamental equation of MD method is Newton's equation of motion for all atoms considered in the system:

$$\ddot{\mathbf{r}}_i = \mathbf{f}_i / m_i, \tag{1}$$

where $\mathbf{r}_i$ and m_i are the position vector and mass of the i-th atom, respectively, and $\mathbf{f}_i$ is the force acting on the i-th atom, which is represented as

$$\mathbf{f}_i = -\partial\Phi / \partial\mathbf{r}_i, \tag{2}$$

with the potential energy Φ of the system considered.
This equation is solved numerically. Verlet's scheme, which is often used in MD simulations, is utilized:

$$\mathbf{r}_i(t + \Delta t) = \mathbf{r}_i(t) + \mathbf{v}_i(t)\Delta t + \mathbf{F}_i(t)\Delta t^2 / (2m_i) \tag{3}$$

$$\mathbf{v}_i(t + \Delta t) = \mathbf{v}_i(t) + (\mathbf{F}_i(t + \Delta t) + \mathbf{F}_i(t))\Delta t / (2m_i), \tag{4}$$

where (t) represents the value at time t, and Δt is the time increment.
Temperature is expressed as

$$T = \frac{2K}{3Nk_b} = \frac{1}{3Nk_b} \sum_i^N m_i v_i^2, \tag{5}$$

where K is the total kinetic energy, k_b is the Boltzmann constant, and the notation $<>$ represents the time average. Temperature is controlled by scaling the velocity with the factor $\sqrt{T/T_0}$. Pressure is defined using the virial theorem, and it can be controlled by adjusting the length of the axes, which is referred to as the scaling method. The stress tensor is defined and controlled employing the so-called Parrinello-Rahman (PR) method (Parrinello & Rahman, 1980,1981).

2.2 EAM potential

Various interatomic energy functions have been proposed and are classified as empirical, semi-empirical, and first-principle potentials. The precision is highest for first-principle potentials, although only a small number of atoms are considered owing to the computational cost. This study employs the EAM potential, which was developed by Daw, Baskes (Daw & Baskes, 1984) and others, and the precision for metals is relatively fine. The potential function is written as

$$\Phi = \sum_i F(\rho_i) + \frac{1}{2} \sum_i \sum_{j \neq i} \phi_{ij}(r_{ij}). \tag{6}$$

Here, Φ is the total potential energy in the system considered, the first term on the right-hand side is a many-body term as a function of the local electron density ρ_i around the i-th atom, and the second term is a two-body term that expresses a repulsive force at close range.
The electron density ρ_i is assumed to be (Clementi & Roetti, 1974)

$$\rho_i = \sum_{j \neq i} \tilde{\rho}(r_{ij}) = \sum_{j \neq i} \{N^s \tilde{\rho}^s(r_{ij}) + N^d \tilde{\rho}^d(r_{ij})\}, \tag{7}$$

where

$$\tilde{\rho}^s(r_{ij}) = \tilde{\rho}^d(r_{ij}) = |\sum_I C_I R_I|^2 / 4\pi, \tag{8}$$

	Z_0	α	β	ν	N^s	N^d	k_1	k_2	k_3
Ni	10	1.554	0.352	0.5	1.7	8.3	-26.883	-34.040	-75.016
Al	3	1.3	1.5	0.5	0.76	2.24	-33.372	43.995	-186.75

Table 1. EAM parameters for Ni and Al in units of eV for energy and Å for length (I).

Ni				Al			
I	n_I	ζ_I	C_I	I	n_I	ζ_I	C_I
s 1	1	54.88885	-0.00389	s 1	1	26.89394	-0.00452
2	1	38.48431	-0.02991	2	1	20.27618	0.08395
3	2	27.42703	-0.03189	3	2	9.45860	-0.11622
4	2	20.88204	0.15289	4	2	6.86435	-0.18811
5	3	10.95707	-0.20048	5	3	3.35342	0.54265
6	3	7.31958	-0.05423	6	3	2.09388	0.55020
7	4	3.9265	0.49292	p 1	2	13.62535	-0.04475
8	4	2.15289	0.61875	2	2	6.90762	-0.14977
d 1	3	12.67582	0.4212	3	3	3.18100	0.26788
2	3	5.43253	0.70658	4	3	1.72743	0.80384

Table 2. EAM parameters (II).

and

$$R_I = \frac{(2\zeta_I)^{n_I+1/2}}{[(2n_I)!]^{1/2}} r_{ij}^{n_I-1} \exp(-\zeta_I r_{ij}). \tag{9}$$

Here, N^s, N^d, C_I, ζ_I, and n_I are parameters that depend on the species of the atom. These parameters are listed in a table by Clementi and Roetti for major metals, and the parameters for Ni and Al in this study are taken from the list; the parameters are given in Tables 1 and 2 in units of eV and Å.

The universal-function method proposed by Rose et al. (Rose et al., 1984) is applied to determine the embedding function F:

$$F(\rho) = k_1\rho^{1/2} + k_2\rho + k_3\rho^2, \tag{10}$$

where k_1, k_2, and k_3 are the parameters for Ni and Al, as given in Table 1.

The second term in Eq. (6) is a two-body term as a function of the distance between two atoms r_{ij}. The following form reported by Rubini and Ballone (Rubini & Ballone, 1993) is used.

$$\phi_{ij}(r_{ij}) = Z_i(r_{ij})Z_j(r_{ij})/r_{ij}\,, \tag{11}$$

$$Z(r_{ij}) = Z_0(1+\beta r_{ij}^{\nu})\exp(-\alpha r_{ij}). \tag{12}$$

Here Z_0, β, ν, and α are the parameters for Ni and Al; they are also listed in Table 1.

3. Model and Conditions

3.1 Simulation Model

Before demonstrating the shape-memory process, the stable structure of Ni-Al alloy ranging from 50% to 75% Ni at various temperatures is investigated using the aforementioned EAM potential. The lattice points of B2 structure are assigned to Ni and Al atoms alternately to make Ni-50%Al alloy. An Al atom is then randomly chosen and replaced by a Ni atom. This procedure is repeated until the designated Ni concentration is reached. Using this model, MD simulations are carried out at a constant temperature under zero pressure with a periodic boundary condition.

Figure 2 presents the initial configuration of Ni and Al atoms and snapshots after a 20000-step calculation for (a) 60%Ni and (b) 68% Ni at 10 K. The crystal structure of the 68% model has apparently different, while no change is observed for the 60% model. The transformed phase is regarded as martensite. The stable structure is found to be B2 or martensite for all Ni concentrations and temperatures, as summarized in Fig. 3. Martensite phase is obtained at low temperature for the high-Ni alloy, while B2 is stable at high temperature in the low-Ni range. Since both structures are obtained in the 64%-70% range, the 68% Ni concentration is used in the following simulations.

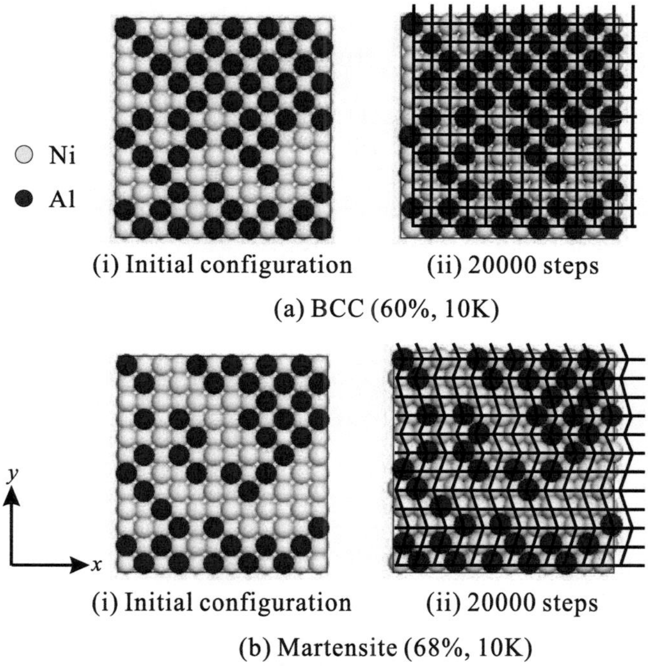

Fig. 2. Initial configuration of atoms and stable structures for (a) 60% Ni and (b) 68% Ni at 10 K.

3.2 Load and temperature condition

Figure 4 represents the external load and temperature profile divided into four stages: loading (I), unloading (II), heating (III) and cooling (IV). The time increment is set as Δt=1.5 fs, and the computation comprises 40000 steps. A shear load is imposed in the loading stage by leaning the edge that runs in the y direction in the x direction at a constant rate. The normal stress component is kept to zero by adjusting the three edge lengths. When the shear strain γ_{xy} reaches 0.33, the stress is released employing the Parrinello-Rahman (PR) method. This stage corresponds to the unloading. The temperature is kept constant at T=10 K through these

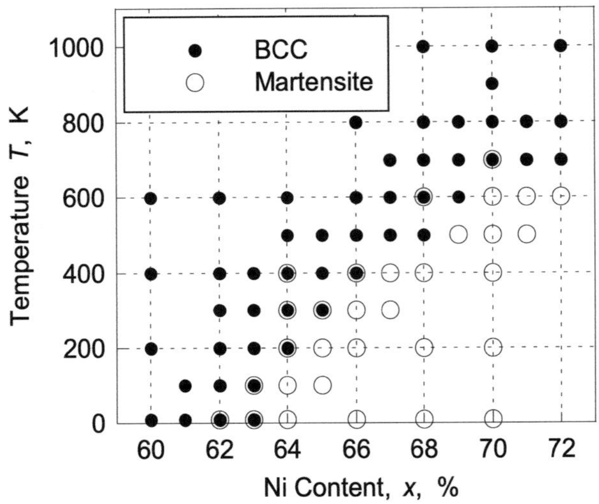

Fig. 3. Phase diagram representing stable structure.

stages, while it is raised to 1000 K and decreased to 10 K in Stages III and IV. Each component of the stress is controlled to zero through heat treatment employing the PR method.

4. Single-Crystal Model

4.1 Deformation and phase transformation

In this section, the results obtained using a single-crystal model with 68% Ni are presented following the work of Uehara et al. (Uehara et al., 2006a). The initial phase is set as martensite

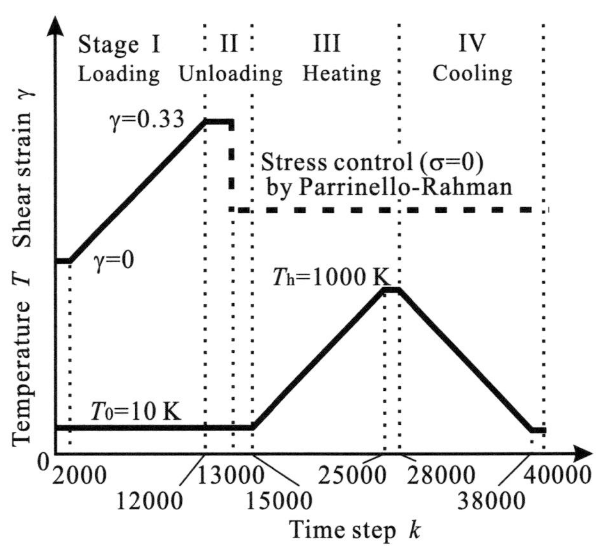

Fig. 4. Temperature and loading profile.

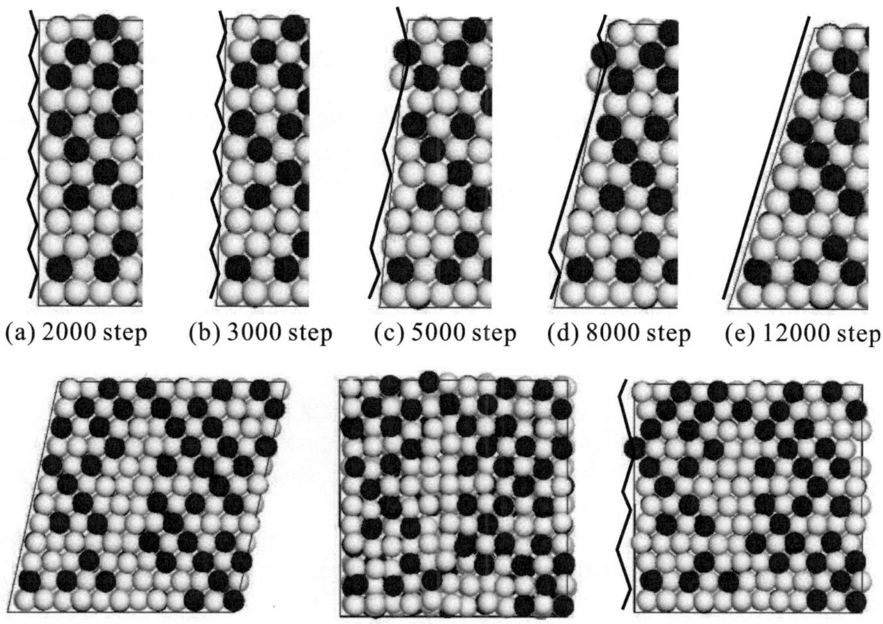

Fig. 5. Configuration of atoms during shear loading (a)–(e), after unloading (f), after heating (g), and in the final state (h).

comprising 864 atoms. Martensite phase in a variant has an lower angle in the $x-y$ plane and is arranged such that layers are parallel to the x-axis. Initially, 12 layers are stacked in the y-direction, and periodic boundary conditions are set in all directions.

Figure 5 represents the variation in the atomic configuration in the loading stage (a)-(e), after unloading (f), after heating (g), and after cooling (h). The Ni and Al atoms are depicted in gray and black, respectively, and auxiliary lines to clarify the variant stacking are drawn in the snapshots of the loading stage. When the shear strain is imposed, the model deforms elastically as shown in Fig. 5(b), in which the orientation of each layer does not change. Some layers change orientation by the 5000th step, and several consecutive layers become uniform, as shown in Fig. 5(c). This change in the layer direction occurs intermittently, and finally a single-variant martensite forms as shown in Fig. 5(e). The system is largely deformed on a macroscopic scale, and the distortion does not return to zero when the external force is released, as shown in Fig. 5(f), except for slight elastic recovery.

In the heating process, there is phase transformation from martensite to the B2 structure, and macroscopic deformation disappears, as shown in Fig. 5(g). Martensite again appears upon cooling. The variant layer is not identical to the initial state, although macroscopically, there is no change in shape, as shown in Fig. 5(h). Therefore, it is concluded that the deformation and shape recovery shown in Fig. 1 are well expressed by this model.

4.2 Stress–strain curve

The stress–strain (S-S) curve for the loading, unloading, heating, and cooling processes, corresponding to the atomistic behavior depicted in Fig. 5, is shown in Fig. 6. The loading curve consists of gradual rises and abrupt drops, and the lines as a whole is zigzag. Each of the stress drops corresponds to the instant that a variant layer changes orientation, and this continues until all layers have the same orientation. The elastic recovery is clearly shown in this figure, and the shape recovery is expressed by the curve returning to the origin. As a result, the S-S curve draws a hysteresis loop, although the loading curve obtained experimentally or macroscopically is much smoother. Nevertheless, if neglecting the zigzag profile in loading, we conclude that the shape recovery is represented.

4.3 Size dependency

To show the size dependency, extensive simulations are carried out using larger models (Uehara et al., 2006b). The S-S curve obtained for a larger model is shown by the dashed line in Fig. 7(a), where the result for a smaller model identical to that in Fig. 6 is also drawn for comparison. The S-S curves have similar tendencies in the sense that there are repeated gradual stress rises and abrupt drops, although the number of peaks and drops is greater for the larger model; i.e., the curve as a whole is smoother for the larger model. The gradients of the two curves in the loading stage are identical, indicating that the elastic modulus is independent of the model size.

Random numbers are used in the preparation of the 68% Ni alloy, and to set the initial velocity of the atoms. Therefore, the results may be affected by the randomness. Figure 7(b) shows the S-S curves obtained in six trials, each of which had the same model size and conditions but different random-number set. The curves do not coincide completely, but mostly show identical tendencies.

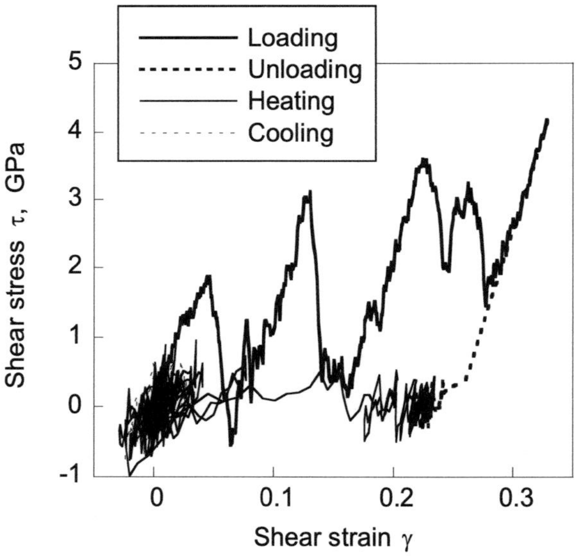

Fig. 6. Stress–strain relation during loading, unloading, heating, and cooling.

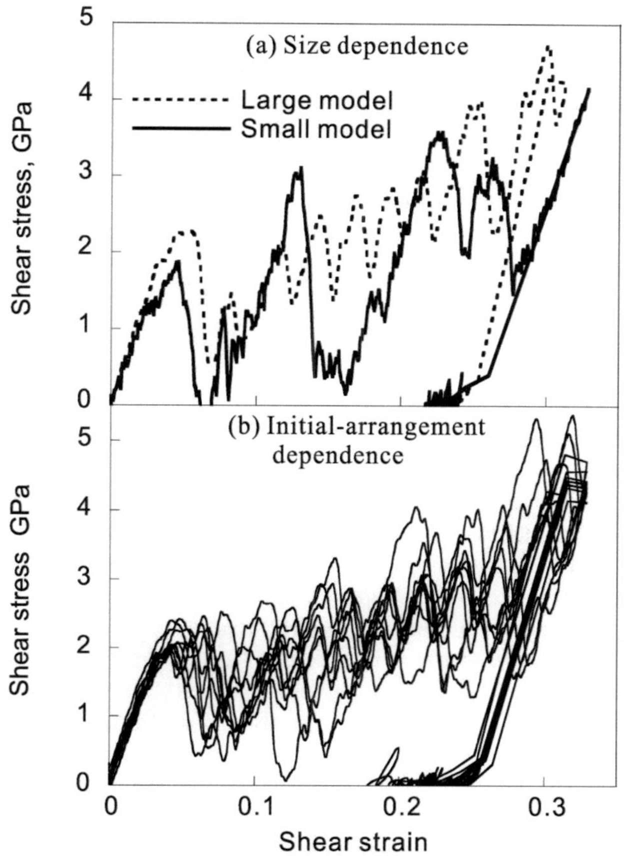

Fig. 7. Stress–strain curves during loading and unloading for larger models.

5. Polycrystalline Model

5.1 Model

Rather large discrepancy in the S-S curve, especially in the loading process, between the MD results and experimental observations is considered to be due to the MD model being highly simplified. To investigate to what degree the model affects the S-S curve, Uehara et al. carried out MD simulations using multi-grain models (Uehara et al., 2008, 2009), and the results are presented in this section.

Two models with different grain shapes and distributions are shown in Fig. 8. Model A consists of two square and two octagonal grains, while Model B has four hexagonal grains. Both models guarantee continuity under periodic boundary conditions. The initial configuration is set as martensite phase, while the orientations of the variants are specific to each grain. The model size is around $40 \times 40 \times 5$ in units of the lattice constant (i.e., about 16.6nm$\times$16.6nm$\times$1.5nm) in the x, y, and z directions, respectively, and the total number of atoms is about 31000. The conditions for shear loading and heat treatment are the same as those in the previous section.

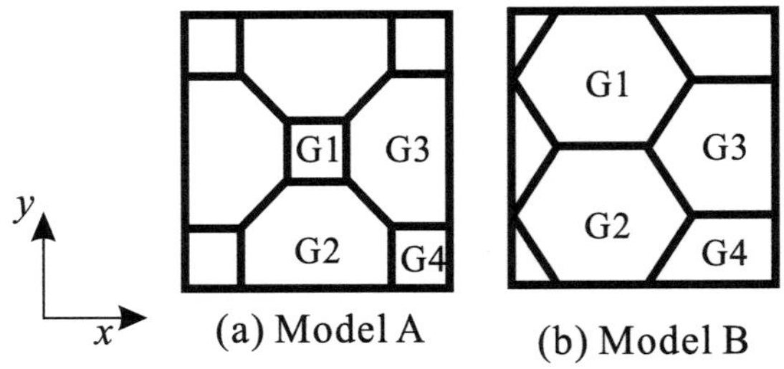

Fig. 8. Schematic illustration of the polycrystalline models.

5.2 Results for Model A

5.2.1 Configuration of atoms

Snapshots of the atomic configuration of Model A are shown in Fig. 9. The color indicates the local structure (Uehara et al., 2009), where M1 and M2 are the martensite phases consisting of variant layers with an alternating sequence and consecutive sequence, respectively.
The initial configuration and local structure are shown in Fig. 9(a). All atoms are arranged at the M1 martensite sites. In the preparation of the multi-grain model, large mismatches are introduced at the grain boundaries. Therefore, some relaxation behavior is observed in the initial stage, resulting in the appearance of B2 and M2 structures around the grain boundaries, as shown in Fig. 9(b). The overall shape of the model is also affected by the relaxation behavior, and slight leaning is observed. However, the changes are slight, and the non-deformed M1 structure remains. Accordingly, the loading simulation is continued after this relaxation stage.
In the loading stage, the model is compulsively deformed; i.e., shear deformation in the x-y plane, as shown in Figs 9(c) and (d). In this stage, M1 atoms turn to M2 phase, and the ratio of M2 atoms increases as macroscopic deformation progresses. A characteristic feature of the M2 appearance is that layers having specific orientation form each grain. This process is mostly identical to that observed for the single-crystal model, but it is notable that the rate of growth of the M2 domain varies among grains, which is considered to be the major cause of the smoothing of the S-S curve as noted later. Finally, transformation from M1 to M2 is mostly complete by the end of loading, as shown in Fig. 9(d), while the grain boundaries (colored in green) remain at their initial positions. The macroscopic deformation disappears with release of the external force, as shown in Fig. 9(e). It is notable that the distribution of the local structure does not change in the unloading process.
In the following heat treatment process, the phase transformation from martensite to B2 occurs similarly to the single-crystal case, as shown in Fig. 9(f), and the macroscopic shape of the model regains its original shape. Note again that the grain boundaries are still distinguishable after B2 transformation. When the model has cooled, martensitic transformation occurs. Since the variants are generated with random orientations, the distribution of M1 and M2 differs from that of the initial state, but this is not critical here. There is in-depth discussion of the deformation mechanism in the literature (Uehara et al., 2009).

5.2.2 Stress–strain curve

Figure 10 represents the stress–strain relation throughout the loading, unloading, and heat treatment processes. The shear strain is added at a constant rate in the loading stage, and the corresponding time steps are shown on the upper side of the diagram. The S-S curve in Fig. 10 dramatically differs from those for single-crystal models shown in Figs. 6 and 7. The zigzag shape in the loading stage disappears and there is a smooth curve instead. The

sudden drop in stress is due to the simultaneous change in the orientation of a specific layer through the model in a single crystal. In the multi-grain model, however, the motion of the layer is interfered by the grain boundary, and the orientation change does not pass thorough the model. Therefore, there is no sudden drop in stress, and other layers begin to deform. This occurs continuously, resulting in the smoothing of the stress variation.
The subsequent elastic recovery in the unloading process — macroscopic shape recovery due to B2 transformation in the heating process — and regeneration of the original martensite phases are similar to what occurs for the single-crystal model, and a hysteresis loop forms. As a result, we conclude that the shape-memory behavior is successfully simulated, and the S-S curves using a multi-grain model approach the experimental curves.

5.3 Results for Model B

Figures 11 and 12 show the variation in the configuration of atoms and the S-S curves in the loading stage obtained using Model B. The snapshots in Figs. 11(a)–(d) correspond to the initial configuration, after relaxation, under loading, and at the end of loading, respectively. In Fig. 12, the S-S curves obtained in six cases, which differ in terms of the variant orientation in each grain, are plotted in a single diagram.
Similarly to what is seen in Fig. 9, there is some relaxation around the grain boundaries, and B2 and M2 domains appear as shown in Fig. 11(b). Here again, since the overall state of M1 phase is maintained, the loading process is continued using this model. The orientation of the martensite varies as the shear load is imposed, and the M2 domain grows. Each variant has a particular orientation within the grain, and finally, M2 occupies almost all grains except at grain boundaries. Note that in some cases, there are multiple orientations in a grain, which is also considered to be the cause of the smoothing of the S-S curve. In Fig. 12, the overall tendency is common for all trials, while the plateau stress is classified into two groups. Detailed discussion is provided in the literature (Uehara et al., 2009).

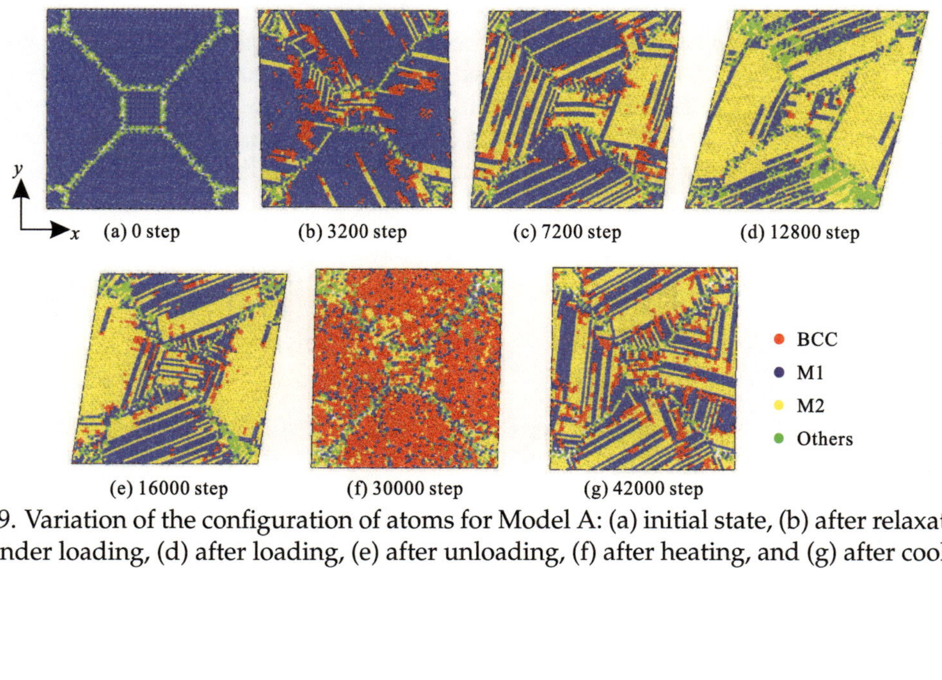

Fig. 9. Variation of the configuration of atoms for Model A: (a) initial state, (b) after relaxation, (c) under loading, (d) after loading, (e) after unloading, (f) after heating, and (g) after cooling.

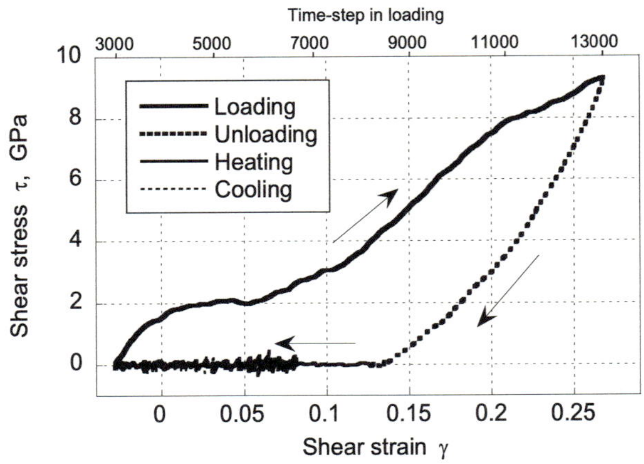

Fig. 10. Stress–strain relation during loading, unloading, heating, and cooling for Model A.

6. Concluding Remarks

Deformation and shape-recovery processes are simulated employing the molecular dynamics method. There is shape memory even in a simple model of a single crystal, and a hysteresis loop for the stress–strain curve is obtained. Deformation of martensite phase progresses through layer-by-layer change in the variant orientation, and the propagation results in a zigzag shape of the S-S curve. The shape of the loading curve drastically changes when using a multi-grain model, and the S-S curve approaches the experimentally observed curve, which is obtained on a macroscopic scale. It is revealed in this study that the smoothing of the curve is due to the existence of grain boundaries and variation in the crystal orientation.

As future work, extensive simulations are required for detailed discussion on the role of the grain boundary and anisotropic tendency. Other effects and mechanisms based on defects, dislocations, sliding, and twinning, which are especially important in SMAs, should be considered. The size dependency is also expected to be investigated in depth in terms of the above-mentioned mechanism, since it is one of the major remaining problems in solid mechanics (Yamakov et al., 2002; Shimokawa et al., 2005). For quantitative evaluation, the precision of the interatomic potential should be definitive and a three-dimensional model used.

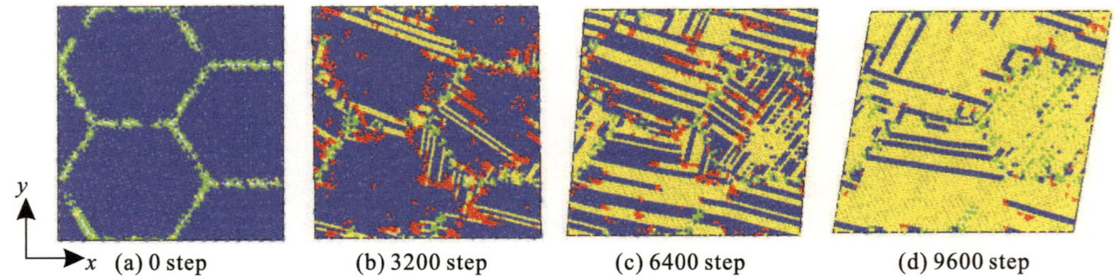

Fig. 11. Variation of the configuration of atoms for Model B: (a) initial state, (b) after relaxation, (c) under loading, and (d) after loading.

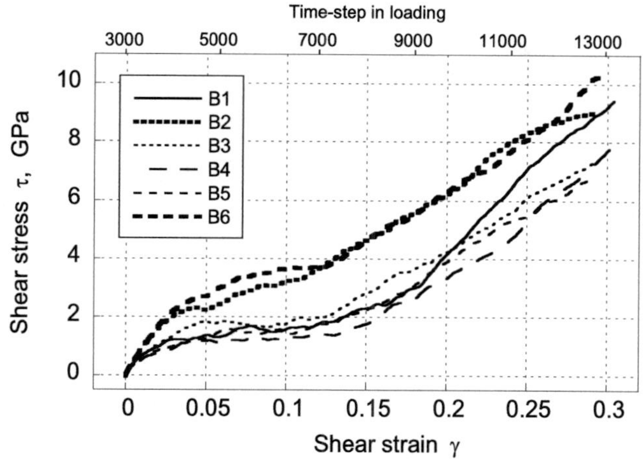

Fig. 12. Stress–strain relation in the loading stage using Model B with different sets of variant orientations in the grains.

Nevertheless, it is concluded that the molecular dynamics simulation is a powerful and promising tool for clarifying the mechanism of the shape-memory behavior and will allow prediction of new functionalities and further development of advanced devices.

7. References

Ackland, G. J.; Jones, A. P. & Noble-Eddy, R. (2008). Molecular dynamics simulations of the martensitic phase transition process. *Mater. Sci. Eng. A*, Vol. 481-482, 11-17.

Chen, S. P.; Srolovitz, D. J. & Voter, A. F. (1989). Computer simulation on surfaces and [001] symmetric tilt grain boundaries in Ni, Al, and Ni_3Al. *J. Mater. Res.*, Vol. 4, 62-77.

Clapp, P. C.; Rifkin, J.; Kenyon, J. & Tanner, L. E. (1988). Computer study of tweed as a precursor to a martensitic transformation of a bcc lattice. *Metall. Trans. A*, Vol. 19, 783-787.

Clementi, E. & Roetti, C. (1974). Roothaan-Hartree-Fock atomic wavefunctions. *Atomic Data and Nuclear Data Tables*, Vol. 14, 177-478.

Daw, M. S. & Baskes, M. I. (1984). Embedded-atom method: Derivation and application to impurities, surfaces, and other defects in metals. *Phys. Rev. B*, Vol. 29, 6443-6463.

Elliott, R. S.; Shaw, J. A. & Triantafyllidis, N. (2002). Stability of thermally-induced martensitic transformations in bi-atomic crystals. *J. Mech. Phys. Solids*, Vol. 50, 2463-2493.

Elliott, R. S.; Shaw, J. A. & Triantafyllidis, N. (2006). Stability of crystalline solids — II: Application to temperature-induced martensitic phase. *J. Mech. Phys. Solids*, Vol. 54, 193-232.

Farkas, D; Mutasa, B.; Vailhe, C. & Ternes, K. (1995). Interatomic potentials for B2 NiAl and martensitic phases. *Modelling Simul. Mater. Sci. Eng.*, Vol. 3, 201-214.

Foiles, S. M.; Baskes, M. I. & Daw, M. S. (1986). Embedded-atom-method functions for the fcc metals Cu, Ag, Au, Ni, Pd, Pt, and their alloys. *Phys. Rev. B*, Vol. 33, 7983-7991.

Foiles, S. M. & Daw, M. S. (1987). Application of the embedded atom method of Ni_3Al. *J. Mater. Res.*, Vol. 2, 5-15.

Huang, X.; Ackland, G. J. & Rabe, K. M. (2003). Crystal structures and shape-memory behaviour of NiTi. *Nature Mater.*, Vol. 2, 307-311.

Ji, C. & Park, H. S. (2007). The effect of defects on the mechanical behavior of silver shape memory nanowires. *J. Comput. Theor. Nanosci.*, Vol. 4, 578-587.

Kastner, O. (2003). Molecular-dynamics of a 2D model of the shape memory effect Part I: Model and simulations. *Continuum Mech. Thermodyn.*, Vol. 15, 487-502.

Kastner, O. (2006). Molecular-dynamics of a 2D model of the shape memory effect Part II: thermodynamics of a small system. *Continuum Mech. Thermodyn.*, Vol. 18, 63-81.

Leo, P. H.; Shield, T. W. & Bruno, O. P. (1993). Transient heat transfer effects on the pseudoelastic behavior of shape-memory wires. *Acta Metall. Mater.*, Vol. 41, 2477-2485.

Ozgen, S. & Adiguzel, O. (2003). Molecular dynamics simulation of diffusionless phase transformation in a quenched NiAl alloy model. *J. Phys. Chem. Solids*, Vol. 64, 459-464.

Ozgen, S. & Adiguzel, O. (2004). Investigation of the thermoelastic phase transformation in a NiAl alloy by molecular dynamics simulation. *J. Phys. Chem. Solids*, Vol. 65, 861-865.

Park, H. S.; Gall, K. & Zimmerman, J. A. (2005). Shape memory and pseudoelasticity in metal nanowires. *Phys. Rev. Lett.*, Vol. 95, 255504.

Park, H. S. & Ji, C. (2006). On the thermomechanical deformation of silver shape memory nanowires. *Acta Mater.*, Vol. 54, 2645-2654.

Parrinello, M. & Rahman, A. (1980). Crystal structure and pair potentials: A molecular-dynamics study. *Phys. Rev. Lett.*, Vol. 45, 1196-1199.

Parrinello, M. & Rahman, A. (1981). Polymorphic transitions in single crystals: A new molecular dynamics method. *J. Appl. Phys.*, Vol. 52, 7182-7190.

Rose, J. H.; Smith, J. R.; Huina, J. & Ferrante, J. (1984). Universal features of the equation of state of metals. *Phys. Rev. B*, Vol. 29, 2963-2969.

Rubini, S. & Ballone, P. (1993). Quasiharmonic and molecular-dynamics study of the martensitic transformation in Ni-Al alloys. *Phys. Rev. B*, Vol. 48, 99-111.

Sato, T.; Saito, K.; Uehara, T. & Shinke, N. (2004). Molecular dynamics study on nano structure and shape-memory property of Ni-Ti alloy. *Trans. Mat. Res. Soc. Japan*, Vol. 29, 3615-3618.

Sato, T.; Saitoh, K. & Shinke, N. (2006). Molecular dynamics study on microscopic mechanism for phase transformation of Ni-Ti alloy. *Modelling Simul. Mater. Sci. Eng.*, Vol. 14, S39-S46.

Shao, Y.; Clapp, P. C. & Rifkin, J. A. (1996). Molecular dynamics simulation of martensitic transformations in NiAl. *Metall. Mater. Trans. A*, Vol. 27, 1477-1489.

Shimokawa, T.; Kinari, T.; Shintaku, S.; Nakatani, A. & Kitagawa, H. (2005). Defect-induced anisotropy in mechanical properties of nanocrystalline metals by molecular dynamics simulations. *Modelling Simul. Mater. Sci. Eng.*, Vol. 13, 1217-1231.

Uehara, T.; Masago, N. & Inoue, T. (2001). An atomistic study on temperature-incorporated phase transformation in Ni-Al alloy. *Proc. 50th JSMS Annual Meet.*, 283-284.

Uehara, T. & Tamai, T. (2004). Molecular dynamics simulations on shape memory effect in Ni-Al alloy. *Proc. 6th World Cong. Comp. Mech.*, CD-ROM.

Uehara, T. & Tamai, T. (2005). Molecular dynamics simulation on shape-memory effect in Ni-Al alloy by using EAM potential. *Trans. Japan Soc. Mech. Eng.*, Vol. 71, No. 705, 717-723. (in Japanese)

Uehara, T. & Tamai, T. (2006a). An atomistic study on shape-memory effect by shear deformation and phase transformation. *Mechanics of Advanced Materials and Structures*, Vol. 13, 197-204.

Uehara, T.; Tamai, T. & Ohno, N. (2006b). Molecular dynamics simulations of the shape-memory behavior based on martensite transformation and shear deformation. *JSME Int. J. A*, Vol. 49, 300-306.

Uehara, T.; Asai, C. & Ohno, N. (2008). Molecular dynamics simulations on the eeformation mechanism of multi-grain shape-memory alloy. *Advances in Heterogeneous Material Mechanics*, Eds. Fan, J. & Chen, H., 316-319.

Uehara, T.; Asai, C. & Ohno, N. (2009). Molecular dynamics simulations of the shape memory behavior using a multi-grain model. *Modelling Simul. Mater. Sci. Eng.*, Vol. 17, 035011.

Wagner, M. F. X. & Windl, W. (2008). Lattice stability, elastic constants and macroscopic moduli of NiTi martensites from first principles. *Acta Mater.*, Vol. 56, 6232-6245.

Yamakov, V.; Wolf, D.; Phillpot, S. R. & Gleiter, H. (2002). Deformation twinning in nanocrystalline Al by molecular-dynamics simulation. *Acta Materialia*, Vol. 50, 5005-5020.

Thermo-mechanical behaviour of NiTi at impact

Zurbitu[a,c], J.; Kustov[b], S.; Zabaleta[a], A.; Cesari[b], E. and Aurrekoetxea[a], J.
[a]*Mechanical and Industrial Production Department, Mondragon Unibertsitatea, Loramendi 4, 20500 Mondragon, Spain*
[b]*Department of Physics, Universitat de les Illes Balears, Cra Valldemossa, km 7.5, 07122, Palma de Mallorca, Spain*
[c]*Mechanical Design Department, Ikerlan-IK4, P°. J. Mª. Arizmendiarrieta 2, 20500 Mondragon, Spain*

1. Introduction

The unique properties of NiTi shape memory alloys have been exploited for engineering applications over the last few decades (Funakubo, 1987; Duerig et al., 1990; Otsuka & Wayman, 1998; Auricchio et al., 2001; Lagoudas, 2008). Recently they have become important for impact applications due to their large recoverable strains and high capacity to dissipate energy. For instance, these alloys are highly attractive for impact damping, seismic protection or health monitoring for impact damage detection (Dolce & Cardone, 2001; Tsoi et al., 2003; Qiu et al., 2006). Nevertheless, the thermo-mechanical behaviour of these alloys at impact, deformation rate in the order of 1 - 10^3 s^{-1}, is not yet well known and only a few works deal with the dynamics of the superelastic transformation at impact strain rates, from the parent phase B2 to the martensitic phase B19′, studying the mechanical behaviour in compression mode (Chen et al., 2001; Xu et al., 2006), in tensile mode (Zurbitu et al., 2009a), or analyzing the dynamics of propagating phase boundaries (Niemczura & Ravi-Chandar, 2006; Zurbitu et al., 2009b). Unfortunately, there is a lack of knowledge on the thermal evolution of NiTi when it is subjected to impact loads, so this chapter will address this issue. This lack of knowledge is mainly due to the absence of a well-established characterization methodology for this regime (Boyce & Crenshaw, 2005). Nevertheless, it has recently been developed a new methodology based on the conventional instrumented tensile-impact test which is able to obtain mechanical properties of NiTi with high accuracy in the impact range (Zurbitu et al., 2009c).
In the present chapter, a complete characterization of the thermo-mechanical behaviour of superelastic NiTi from low to impact strain rates (10^{-4} -10^2 s^{-1}) is presented. In order to reach this objective, the stress-strain response was simultaneously registered with thermographic observations at different strain rates up to impact, in order to link the evolution of the temperature with the mechanical behaviour. Moreover other conventional techniques such as screw-driven testing machines have also been employed to extend the thermo-mechanical knowledge at lower strain rates.

1.1 Overview of strain rate effect on the mechanical properties of NiTi

It is well known that the stress-induced martensitic transformation is exothermic, whereas the reverse transformation is endothermic. Characteristic stresses and strains of these transformations depend on the temperature, and since the strain rate could change the heat-transfer phenomena, the temperature could vary during the deformation modifying the mechanical response of NiTi (Shaw & Kyriakides, 1995), mainly in terms of transformation stresses or hysteretic behaviour, Fig. 1. The particular interaction between the strain rate, temperature and transformation stress, enhances the sensitivity of the mechanical behaviour to the strain rate, which is a key parameter in the design of impact applications and must be accurately taken into account. So, the knowledge of the strain-rate effects on the thermo-mechanical properties of NiTi is necessary and crucial for the design and optimization of impact applications.

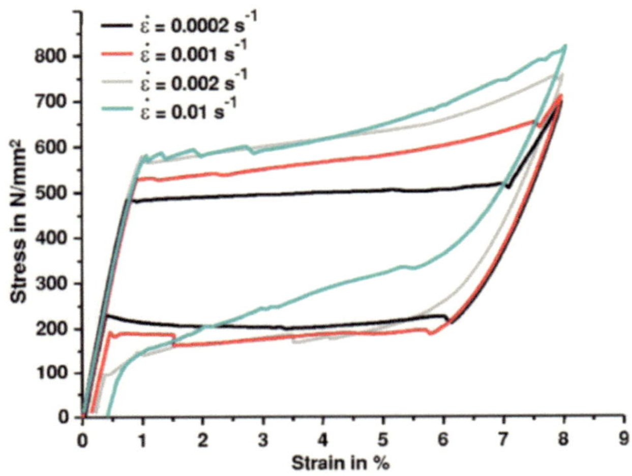

Fig. 1. Stress-strain curves of superelastic NiTi wires with a diameter of 0.9 mm under tension at various strain-rates. [After (Schmidt, 2006)].

Below certain limit, the mechanical behaviour of NiTi is strain rate independent. This is on the order of 10^{-4} s^{-1} (Shaw & Kyriakides, 1995; Schmidt, 2006) and it may be considered as the strain rate limit below which there is enough time to allow all the transformation heat to be completely exchanged with the surroundings. As a result of this feature the temperature does not change in the specimens and the deformation process may be considered as an isothermal one, keeping invariable the mechanical behaviour and constant the transformations stresses.

As the strain rate is raised from the quasi-static limit, the forward transformation stresses increase while the reverse transformation stresses decrease widen then the hysteresis, Fig. 1, as is widely corroborated in the literature (Shaw & Kyriakides, 1995; Vitiello et al., 2005; Schmidt, 2006; Zurbitu et al., 2009a). Above certain strain rate this trend changes, Fig. 2. Thus, the forward transformation stresses stop increasing and become constant after 10^{-1} s^{-1}, and the reverse transformation stresses change their tendency and increase for strain rates higher than 10^{-3} - 10^{-2} s^{-1} (Tobushi et al., 1998; Vitiello et al., 2005; Pieczyska et al., 2006a; Schmidt, 2006; Zurbitu et al., 2009a). The combination of these factors narrows the hysteresis. At high strain rates, the time for heat exchange between the surroundings and

the specimen is reduced so part of the transformation heat is spent in heating up or cooling down the specimen changing the transformation stresses.
It is known that this trend continues until the impact range, **10^{-2}** s^{-1} (Zurbitu et al., 2009a), but it is unknown the temperature evolution at those high strain rates. At impact, the time for heat exchange is drastically reduced and the deformation process is closer to adiabatic conditions, so it is necessary a deeper understanding of the adiabatic nature of the stress induced transformation on the thermo-mechanical behaviour of NiTi.

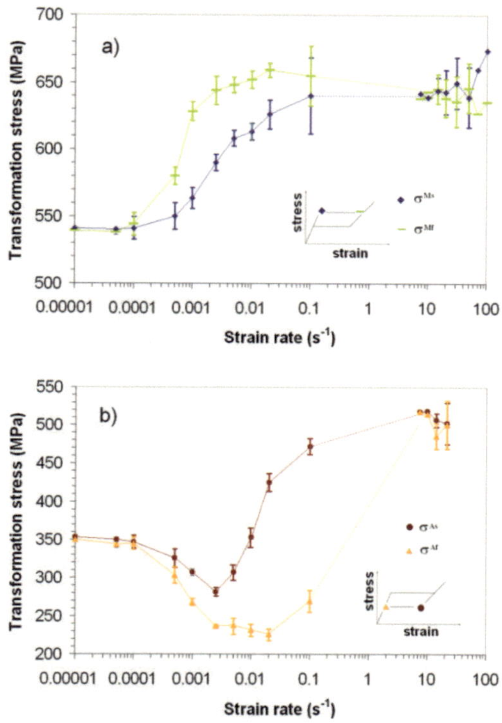

Fig. 2. Variation of the transformation stresses with strain rate: (a) forward SIM transformation stresses, and (b) reverse SIM transformation stresses for a load-unload cycle with complete SIM transformation. [After (Zurbitu et al., 2009a)].

Most of works that study the detwinning of the stress induced martensite in NiTi wires as a function of the strain rate in the tensile configuration, cover the low strain rate range (10^{-5} - 1 s^{-1}) using conventional screw-driven or servo-hydraulic testing machines (Shaw & Kyriakides, 1995; Lin et al., 1996; Wu et al., 1996; Tobushi et al., 1998; Entermeyer et al., 2000; Prahlad & Chopra, 2000; Vitiello et al., 2005; Pieczyska et al., 2006a; Schmidt, 2006). Only a few works deal with the response at very high strain rates in the ballistic range (< 10^3 s^{-1}), employing the Split Hopkinson Pressure Bar (SHPB) (Miller et al., 2000; Adharapurapu et al., 2006; Bragov et al., 2006). The characterization in the intermediate range (1 - 10^3 s^{-1}), has been recently possible (Zurbitu et al., 2009a) thanks to the improvement of the conventional instrumented tensile-impact test which is able to measure with high accuracy mechanical properties at impact strain rates (Zurbitu et al., 2009c). Regarding the thermal evolution as a function of the strain rate it is limited to the low range (10^{-5} - 10^{-1} s^{-1}) (Chrysochoos et al., 1995; Li et al., 1996; Chang et al., 2006; Pieczyska et al., 2007), which offers the chance to study the thermo-mechanical behaviour of NiTi at impact strain rates using the new set up of the conventional instrumented tensile-impact test in combination with thermographic techniques, Fig. 3.

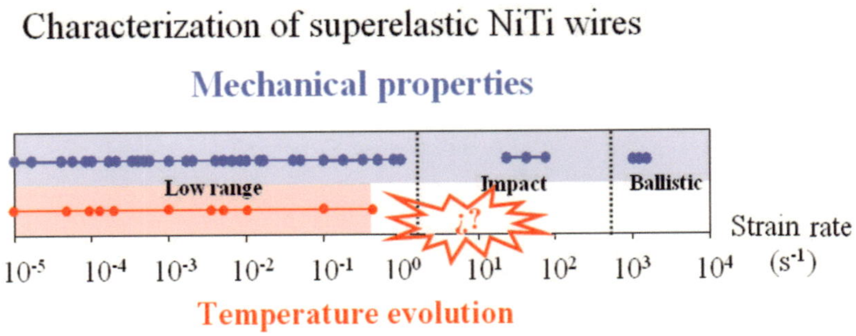

Fig. 3. Overview of the characterization of superelastic NiTi wires as a function of the strain rate

2. Measurement of properties at impact strain rates

2.1 Material: Superelastic NiTi wire

For the experimental work carried out and presented in this chapter it has been chosen a NiTi alloy in the form of wire, 0.5 mm in diameter, showing superelastic behaviour at room temperature, which is commonly used for many researches and applications. Specifically this is a commercially available NiTi wire (ref. NT09), with a nominal composition of 50.9 at.% Ni balanced with Ti, purchased from AMT. The material was supplied in the form of cold drawn wire with a sandblasted surface condition and an average cold worked for 45% based on area reduction with a continuous straight annealing heat treatment at 520 ºC for 30 s in order to optimize the superelastic behaviour. The transformation temperatures deduced from the DSC (Differential Scanning Calorimeter) curves obtained at heating/cooling rates of 10 K/min in the fully annealed condition show that B19' martensite may be induced by tension at room temperature. (M_f = -51.9 ºC; M_s = -31.4 ºC; A_s = -24.6 ºC; A_f = -6.4 ºC). More detailed information about the material may be found in ref. (Zurbitu et al., 2009a).

2.2 Experimental technique: Stress-Strain-Temperature measurements

In order to obtain the thermo-mechanical properties at impact strain rates some measurements of the stress-strain response and the temperature evolution were taken simultaneously as shows the specially developed experimental set-up of the Fig. 4. This is based on an improved instrumented tensile-impact device (Zurbitu et al., 2009c) which is able to obtain stress-strain measurements with high accuracy at high strain rates, on the order 1-102 s-1. It consists on an impactor which deforms the sample by hitting a mobile grip at which the sample is attached. The impact force is measured by a piezoelectric sensor ICP® quartz force ring attached to the other grip which is fixed. The stress may be easily calculated by dividing the total force by the section of the specimen. The measurement of the strain during the impact was obtained by the integration of the velocity profile of the mobile grip divided by the initial length of the sample. Velocity measurements were carried out with a laser-based non-contacting equipment POLYTEC OFV-505. For the temperature measurements, infrared thermographic pictures were taken at a frame rate of 1250 Hz with a high speed thermographic camera Flir Titanium 550M during the tensile deformation of the NiTi wires. So, during deformation, simultaneous measurements of infrared radiation, force and velocity were performed so that the temperature is known and the transformation fronts may be visible while it is known the stress-stain state. For the correlation between the transformation stresses and the temperature it is necessary to know not only the emissivity coefficient of the NiTi but also the stress-temperature phase diagram, elements which are detailed below. For the stress-strain-temperature correlation at lower strain rates (10-4 - 10-1 s-1), tests were carried out in conventional screw-driven testing machines INSTRON 4206 and ZWIK Z100 with simultaneous measurement of temperature using the above described thermographic camera.

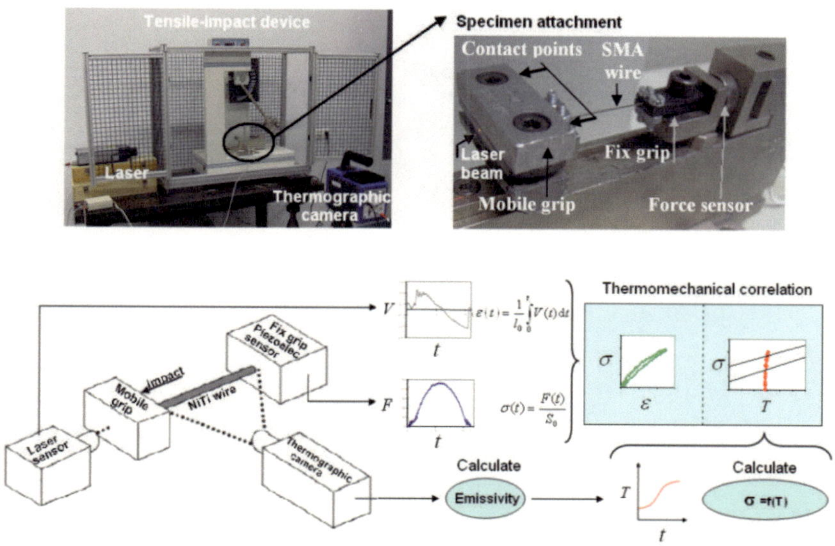

Fig. 4. Experimental set-up for the thermo-mechanical characterization at impact strain rates.

2.3 Emissivity coefficient of NiTi

In the reviewed literature it is possible to find different values of the emissivity coefficient of NiTi, from 0.66 (Iadicola & Shaw, 2007) to 0.83 (Shaw & Kyriakides, 1997). Nevertheless, this value is strongly dependent not only on environmental factors such as the temperature, but also on factors inherent to the material itself such as the composition, roughness, geometry, etc., and therefore it must be calculated for the given testing conditions and the specific material used. Here, for the calculation of the emissivity coefficient of NiTi, samples were employed as described in material section, mounted in the experimental set-up shown in Fig. 3 following the next procedure. On one hand it was measured the temperature of the NiTi wire on the surface by a K-type thermocouple welded to the sample with a high thermal conductivity resin Omegatherm®201. On the other hand a simultaneous thermographic picture of the same area was obtained by means of the thermographic camera described above, and the emissivity value was adjusted until both temperatures match each other. This procedure was repeated for different specimen temperatures achieved by direct Joule effect by passing electric current through the wire ranging from 0.3 to 2A at 0.1 A intervals. The emissivity coefficient that better fits the temperature of the thermographic data with the thermocouple temperature, within the range from room temperature to 200 ºC, is 0.74, Fig. 5. This emissivity value may be considered relatively high and is due to the roughness of the sandblasted surface finish condition and the superficial oxide layer which increases the emissivity. High values of emissivity minimize the scattering due to the reflections generated by heat sources close to the sample when measuring near room temperatures. It is worth to mention that a constant emissivity coefficient can be only considered under the assumption of a grey body which means that this constant value is only valid for measurements carried out within a certain wavelength interval, limited by the resolution of the thermographic camera which in this case ranges from 3.5 to 5.1 μm.

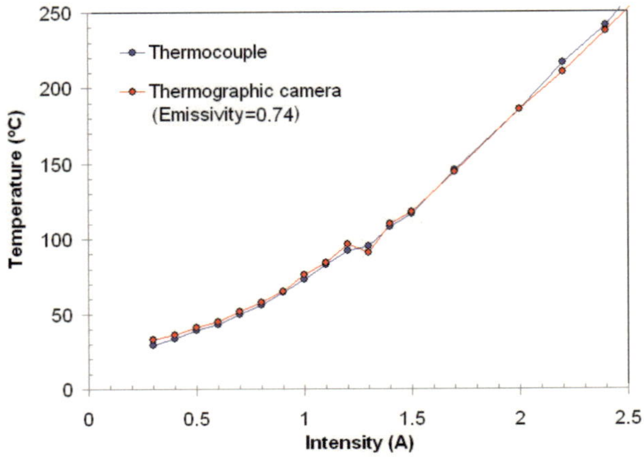

Fig. 5. Adjustment of the emissivity of a NiTi wire with a sandblasted surface condition.

2.4 Stress-temperature phase diagram

By means of the phase diagram shown in Fig. 6(a) it is possible to determine the phase of the material as function of the stress and temperature. This diagram may be used for the thermo-mechanical correlation between the properties obtained from tensile tests, such as the transformation stresses, and the evolution of the temperature. The phase diagram was built based on stress-strain tests with complete stress induced martensitic transformation at different temperatures, Fig. 6(b). The transformation stresses were obtained from these tests at 3% in strain during the forward transformation, and at 2.5% during the reverse one. These tests were carried out in a conventional screw-driven testing machine equipped with climatic chamber. For the correct determination of the phase diagram it is necessary to ensure that the specimen temperature is homogeneous and the same than that kept by the climatic chamber. For this, the tests were conducted under strain controlled conditions at a uniform strain rate of 10^{-4} s^{-1}, low enough to ensure that all the transformation heat is removed from the sample to the surroundings and grips keeping the temperature in the sample stable. The temperature dependence of transformation stresses is given by the Clausius-Clapeyron equation 1, where C_M (forward transformation) and C_A (reverse transformation) are constants and calculated by linear regression as 5.9 and 6.7 MPa·K^{-1} respectively for the alloy here employed.

$$\frac{d\sigma_M}{dT} = C_M ; \frac{d\sigma_A}{dT} = C_A \qquad (1)$$

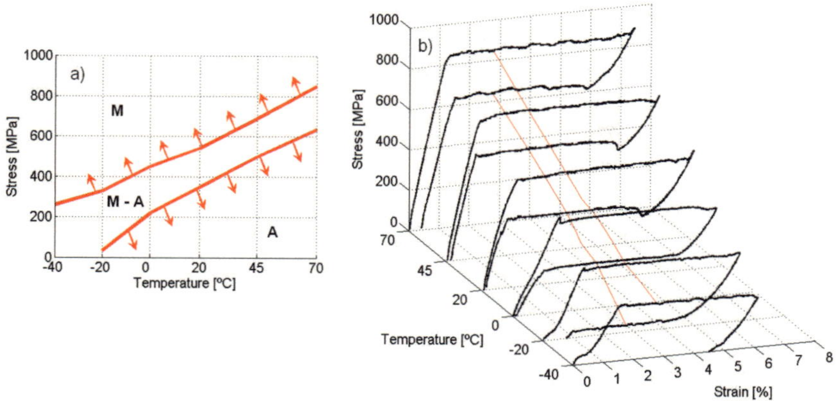

Fig. 6. Stress-temperature phase diagram of NiTi (a), obtained from stress-strain tests with complete stress induced martensitic transformation at different temperatures (b).

3. Thermo-mechanical behaviour of NiTi at impact strain rates

Thanks to the experimental set-up shown in Fig. 4, the determination of the emissivity coefficient of the NiTi and the stress-temperature phase diagram, it is possible to obtain the thermo-mechanical response of NiTi at high strain rates. As a summary of the results obtained, these have been divided into three basically different types of behaviour, a) elastic deformation of the austenitic phase, b) stress induced martensitic transformation and c) elasto-plastic deformation of the martensitic phase.

3.1 Elastic deformation of the austenitic phase at impact

In this case NiTi wire specimens as described above, of 79 mm in length were used. Tests were carried out with an impactor mass of 1.098 kg at an impact velocity of 0.35 m/s corresponding to a strain rate of 4.4 s^{-1}. The force-time graph shown in Fig. 7(a) suggests an elastic deformation of the austenitic phase, which is supported by the stress-strain curve of the Fig. 7(d). Regarding the strain-rate evolution during the impact tests, Fig. 7(d), it keeps close to the initial one during most part of the test and only differs considerably at the highest strain, where the velocity must pass by zero in order to perform the unloading. The very small hysteretic loop observed in the stress-strain curve of the Fig. 7(d) could be due in part to the rearrangement of a certain amount of R-phase variants transformed from the austenitic phase at lower stresses than the stress-induced B19′ martensitic phase (Zurbitu et al., 2009a). The austenite/R-phase transformation is almost negligible in terms of dissipated energy when it is compared with the stress-induced B19′ transformation, but it is visible in Fig. 7(d) as a small slope variation around 0.5% in strain and a small hysteresis. Moreover, during the deformation, small defects may arise even at the elastic range due to the high strain rates. This may explain the small plastic strain after the test shown in Fig. 7(d). Regarding the evolution of temperature, Fig. 7(c), it increases slightly during the loading path because of the small heat generated during the parent phase/R-phase transformation that cannot be released from the specimen due to the high strain rate. During the unloading path, the temperature decreases due to the endothermic character of the retransformation to the austenitic phase, but the small heat generated by defects is kept in the material so the temperature-time curve is not symmetrical and is slightly offset, Fig. 7(c). Comparing the stress and temperature evolution together with the phase diagram, Fig. 7(e), it is shown that the stress just reaches that necessary to induce the martensitic transformation but it does not transform.

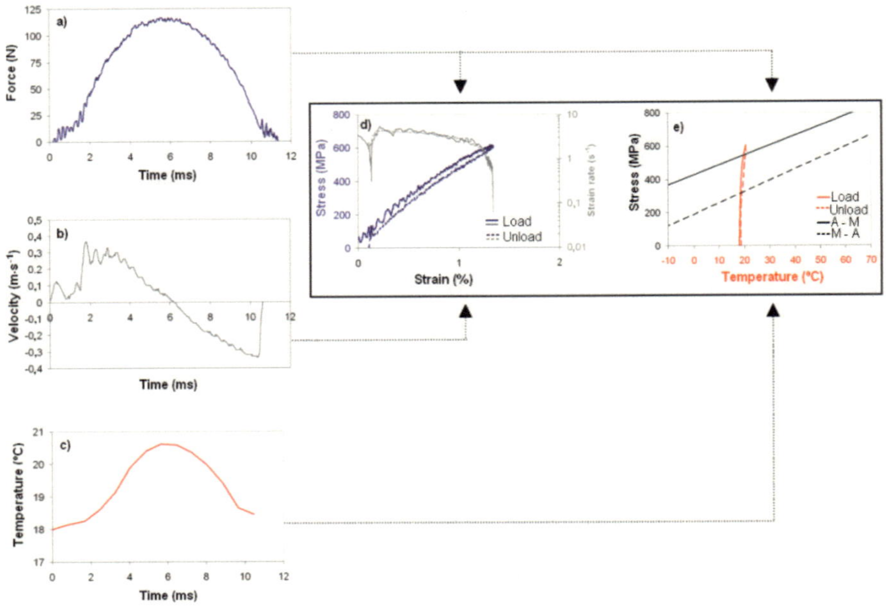

Fig. 7. Elastic deformation of the austenitic phase of NiTi at impact strain rates, a) force-time, b) velocity-time, c) temperature-time, d) stress-strain and strain rate, e) stress-temperature.

3.2 Stress induced martensitic transformation at impact

Here the impactor mass and the initial length is the same than in the previous case, but the impact velocity increases to 1.21 m/s corresponding to a strain rate of 15 s^{-1}. The experimental results show that at impact strain rates, it is still possible to induce the martensitic transformation in NiTi wires, Fig. 8(a). When the force is high enough (b_1), the austenitic lattice becomes thermodynamically unstable being the energy necessary to induce the martensitic transformation lower than that necessary to continue with the elastic deformation of the martensitic phase. Here is shown that at impact the detwinning process occurs at a constant force (b_1 - b_2) as is shown in (Zurbitu et al., 2009a), similarly than that observed at very low strain rates (Shaw & Kyriakides, 1995). From the point (b_2) almost all the specimen consist on detwinned martensite and the strain goes on with the elastic deformation of this phase increasing the force. Once the maximum force is achieved (b_3), the elastic unloading of the martensitic phase begins up to (b_4). Here the martensitic phase becomes unstable and the material transforms back to the parent phase also at a constant force (b_4-b_5). At (b_5) all the martensitic phase is transformed into austenite and the elastic unloading of the austenitic phase takes place. Once the force is removed, most of the deformation is recovered but it may be observed a small portion of permanent deformation, Fig. 8(d), due to the increment of dislocation density and slips occurred during deformation of the martensitic phase. Similarly to that observed at very low strain rates (Liu et al., 2002), the slope of the beginning of the elastic deformation of the martensitic phase is smooth since there is a portion of residual austenite which needs a continuous increment of force to be transformed. The elastic modulus of martensite is higher during unloading than during loading. This is due to the differences in deformation mechanisms. During unloading, the elastic recovery of the martensitic phase prevails, while during loading, not only elastic deformation of martensite but also strain hardening and residual transformation of austenite occurs.

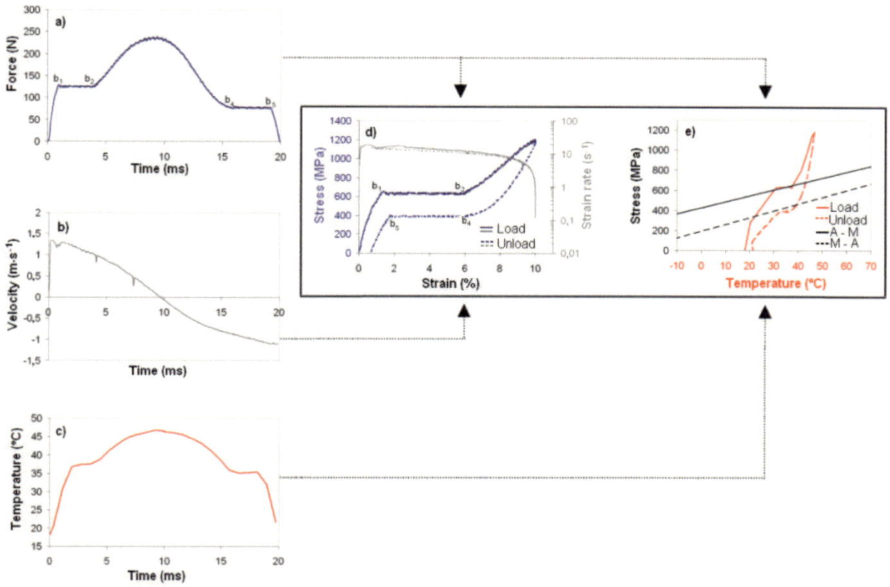

Fig. 8. Stress induced martensitic transformation at impact strain rates, a) force-time, b) velocity-time, c) temperature-time, d) stress-strain and strain rate, e) stress-temperature.

During the transformation, the temperature in the martensitic phase may be up to 20°C higher than the initial one, Fig. 8(c). As a result of this feature, the transformation stresses are higher at impact than at quasi-static strain rates, where the temperature remains unchanged during the deformation, Fig. 9. This is due to the inherent sensibility of the characteristic transformation stresses to the temperature in NiTi, as establishes the Clausius-Clapeyron relationship shown in equation 1. During the deformation at very low strain rates, the transformation heat may be removed from the sample to the surroundings and grips, and the temperature keeps constant during the transformation. In this case, the deformation may be considered as an isothermal process. When the strain rate increases up to impact levels, the time necessary to remove the transformation heat is so reduced that the deformation process may be considered closer to the adiabatic conditions, and the most part of the heat generated during the forward exothermic transformation is spent in warming up the sample raising the transformation stresses.

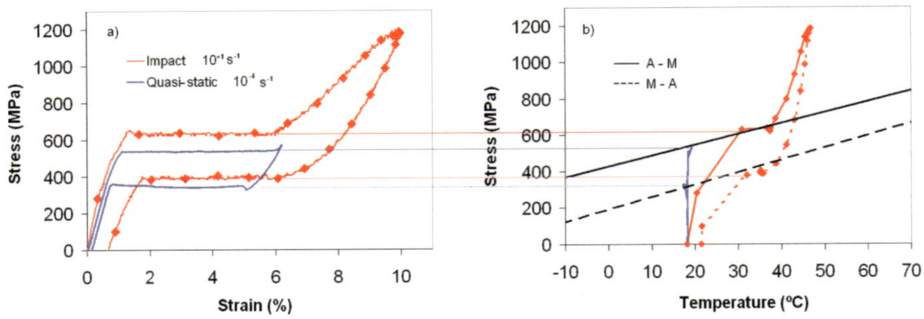

Fig. 9. Correlation between the transformation stresses and temperature stresses for different strain rates.

In the theoretical case of fully adiabatic deformation, the rise in temperature ΔT should be 29.1°C, equation 2, where C_e is the specific heat of NiTi (data provided by the manufacturer) 490 J kg^{-1} K^{-1}, and $\Delta H_{A\text{-}M}$ the transformation enthalpy, 14.25 kJ kg^{-1}, obtained from measurements at a heating/cooling rate of 10 K/min. Comparing the theoretical temperature variation with the experimental one, it is observed that, in fact, the deformation process at impact strain rates is close to adiabatic conditions.

$$\Delta T = \frac{\Delta H_{A-M}}{c_e} \tag{2}$$

According to the value obtained for the sensibility of the forward transformation stresses with the temperature applying the Clausius-Clapeyron relationship, C_A=5.9 MPa·K^{-1}, the theoretical stress increment corresponding to a 20°C variation is 114 MPa. This value is very close to the experimental measurements obtained from tests like those of the Fig. 9, which is 120 ± 10 MPa.

3.3 Elasto-plastic deformation and failure of the martensitic phase at impact

Increasing the impact energy, the elastic deformation of the stress induced martensitic phase continues up to the failure instead of the unloading, Fig. 10. In this case the impactor mass is the same than in the previous cases but the impact velocity **is** raised up to 1.63 m/s. The sample initial length was 31 mm so the corresponding strain rate was 53 s^{-1}. Following the transformation at constant force, Fig. 10(a) (d_1–d_2), the strain continues with the elastic deformation of the martensitic phase (d_2–d_4) together with the residual transformation of austenite (d_2–d_3) as is shown in the previous section. In (d_4–d_5) the stress is so high that the dislocation density and slips increase as is evidenced by the strain-hardening shown in the stress-strain diagram of the Fig. 10(d). Finally, the stress decrease from point d_5 suggests necking before failure. For this higher strain rates, on the order of 10^2 s^{-1}, both the rise in temperature during the transformation (18°C), Fig. 10(c), and the stress increment respect to the quasi-static case (125 MPa), are similar that at lower impact strain rates, 1-10 s^{-1}. So in the strain rate range studied the fact of an increase of the strain rate does not involve further increases in temperature or stresses during the transformation. This means that once the deformation process reaches the quasi-adiabatic condition, the efficiency of the transformation heat in warming the sample reaches the maximum. So, it may be considered that further increments of strain rate will not cause further increases of temperature or transformation stresses; at least while the deformation mechanism remains unchanged.

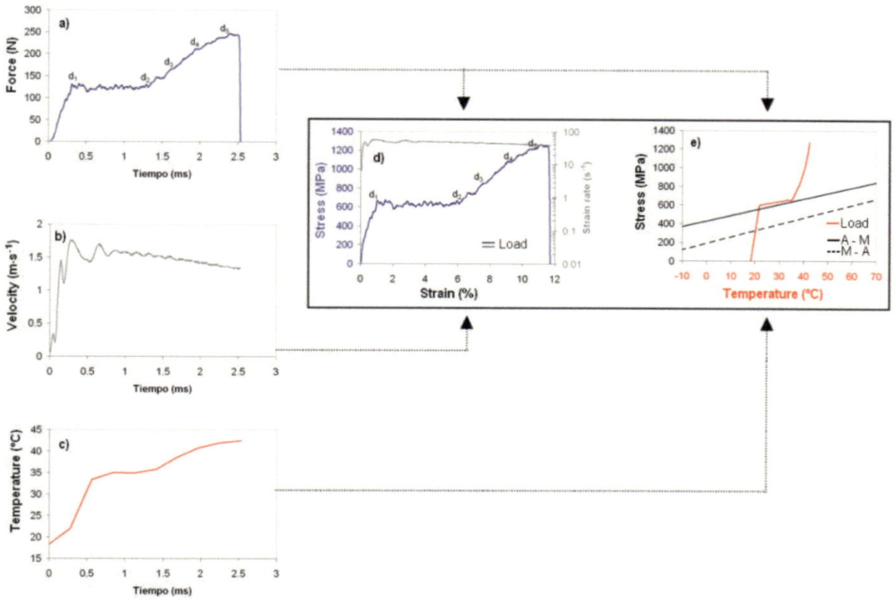

Fig. 10. Elasto plastic deformation of the stress induced martensitic phase at impact strain rates, a) force-time, b) velocity-time, c) temperature-time and strain rate, d) stress-strain and strain rate, e) stress-temperature.

4. Thermo-mechanical behaviour of NiTi at lower strain rates

Below the isothermal strain rate limit, on the order of 10^{-4} s^{-1} for NiTi, the temperature remains unchanged so that the transformation stresses do not vary during deformation as is shown in Fig. 9 and corroborated by other authors (Shaw & Kyriakides, 1995). At higher strain rates, the forward transformation stresses, σ^{Ms} and σ^{Mf}, increase due to the rise in temperature, while the reverse ones, σ^{Ms} and σ^{Mf}, decrease due to the lower temperatures reached in the sample, Fig. 11. The reason of this behaviour is widely supported in the literature (Shaw & Kyriakides, 1995; Wu et al., 1996; Tobushi et al., 1998; Liu et al., 2002). When the strain rate increases, the time necessary to allow the transformation heat exchange with the surroundings is reduced. Thus, a fraction of the heat generated during the exothermic forward transformation remains in the sample increasing **its** temperature and hence the transformation stresses. In the same way, during the endothermic reverse transformation, part of the heat absorbed is transferred from the sample diminishing **its** temperature and the characteristic transformation stresses.

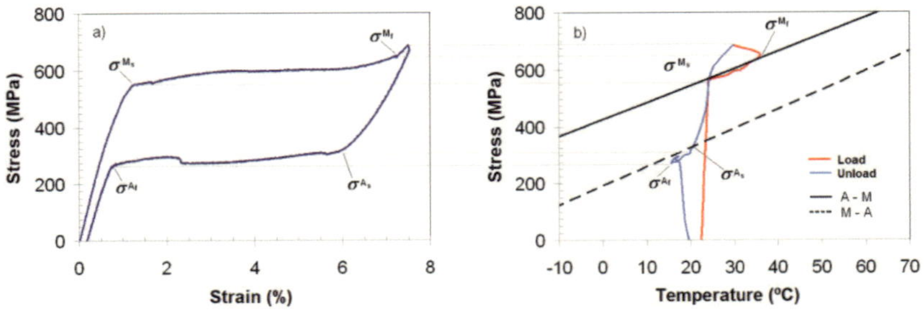

Fig. 11. Thermo-mechanical behaviour of NiTi deformed at a strain rate of 10^{-3} s^{-1}, a) stress-strain diagram, b) temperature evolution of the transformation zone.

The higher the strain rate is, the greater this effect is, Fig. 12, since the time for heat exchange is reduced even further. The stress variation is more pronounced at the end than at the beginning of the transformations because the temperature of the transformation zone increases as the transformation progresses. This causes an increase in the slope of the stress-strain diagram. On the other hand, as the strain rate increases it becomes more evident the residual transformation zone which requires an increase in the external load to be transformed, Fig. 12. The temperature evolution shown in these zones suggests that the residual transformation really occurs, because the temperature continues increasing or decreasing during the residual transformations while during the elastic deformation the evolution of the temperature changes since the heat input/absorption of heat stops. It is worth mentioning that the small stress peaks shown during transformations in the stress-strain diagram of the Fig. 12 are related with new nucleations of the martensitic phase. This aspect is further discussed in the next section.

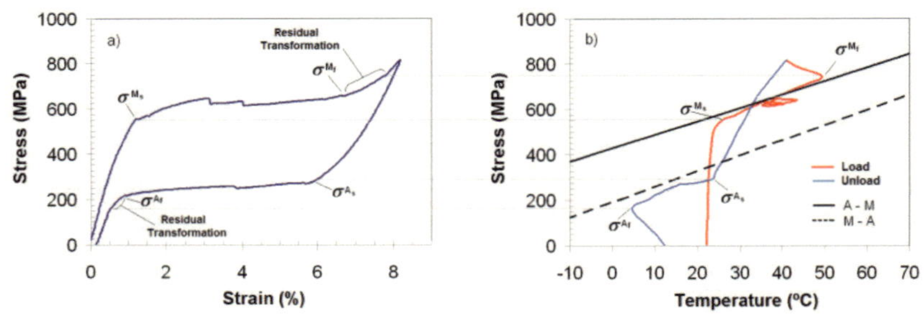

Fig. 12. Thermo-mechanical behaviour of NiTi deformed at a strain rate of $5x10^{-3}$ s^{-1}, a) stress-strain diagram, b) temperature evolution of the transformation zone.

It is known that there is a strain rate value, around 10^{-1} s^{-1} (Zurbitu et al., 2009b), above which this parameter seems to have little influence over the forward transformation stresses. Here is shown that at this strain rate, the temperature reached during the forward transformation, Fig. 13(b), is similar to that observed at impact, with strain rates two orders of magnitude higher, Fig. 9(b). Thus, above 10^{-1} s^{-1}, the deformation process during loading may be considered close to adiabatic conditions and the forward transformation stresses are not influenced by strain rate, Fig. 14.

Regarding the reverse transformation, it is also well known that characteristic transformation stresses change their decreasing tendency and rise above certain strain rate located around 10^{-3} - 10^{-2} s^{-1} (Schmidt, 2006; Zurbitu et al., 2009b). Here is shown that this is due to the higher temperatures developed during the unloading, Fig. 13(b). While at lower strain rates there is enough time to cool the specimen during the elastic recovery of the martensitic phase up to the initial temperature, Fig. 11 and Fig. 12, at 10^{-1} s^{-1}, the time is so reduced that the temperature at the beginning of the reverse transformation is clearly higher than the initial one so the stresses are higher too, Fig. 13. At this strain rate, the temperature along the reverse transformation is reduced decreasing thus the stress. Nevertheless, at impact, the reverse transformation stress keeps constant along the whole reverse transformation, Fig. 14, since the temperature keeps also constant during the process, Fig. 9(b).

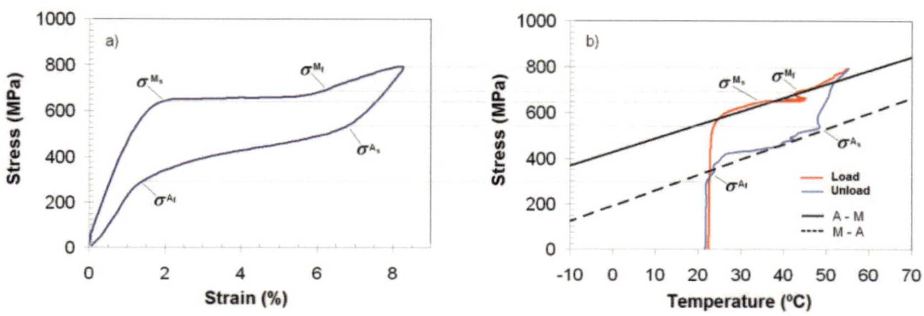

Fig. 13. Thermo-mechanical behaviour of NiTi deformed at a strain rate of 10^{-1} s^{-1}, a) stress-strain diagram, b) temperature evolution of the transformation zone.

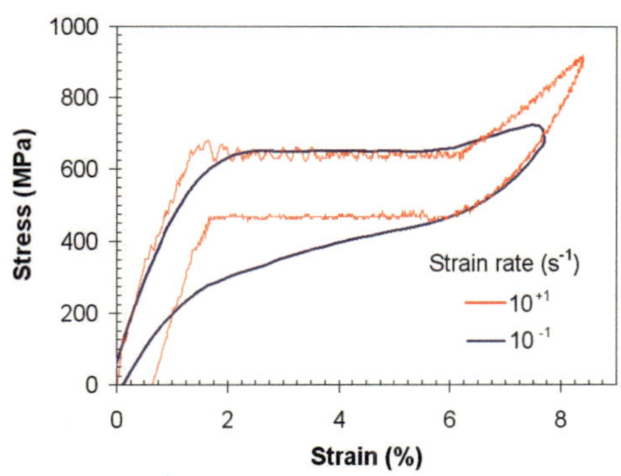

Fig. 14. Stress-strain behaviour of NiTi at different strain rates.

5. Nucleation and phase transformation front evolution

The martensitic transformation occurs by the nucleation and propagation of phase transformation fronts. This inhomogeneous deformation mode divides a deformed sample into transformed and non-transformed zones; and is in this interface, the phase transformation front, at which the lattice distortion takes place. It is well known that the evolution of the amount of transformation fronts strongly depends on the strain rate, at least for the low strain rate range (Leo et al., 1993; Shaw & Kyriakides, 1995), nevertheless the dynamics of the transformation is unknown when it is induced at impact strain rates.

The progress of phase transformation fronts may be observed in many different ways, e.g. by means of strain gauge measurements at different locations (Shaw & Kyriakides, 1995), or measuring changes in temperature along the sample length. The exothermic/endothermic character of the forward/reverse SIM transformation changes the local temperature making visible the nucleation and propagation of phase transformation fronts. This information may be obtained in several ways, by means of small contact thermocouples (Shaw & Kyriakides, 1995), or by means of thermographic pictures (Shaw & Kyriakides, 1997). The latter technique is able to capture the temperature of a greater number of points, so that the resolution in regard to the nucleation and evolution of the phase transformation fronts is higher.
In any of cases, the maximum strain rate at which these tests may be carried out, is directly related to the maximum sampling rate of the measurement systems. In the most recent works, the maximum strain rate at which phase transformation fronts have been observed via thermographic pictures is on the order of 10^{-1} s^{-1} (Pieczyska et al., 2006b), and only a few groups have studied the dynamics of the martensitic transformation at impact monitoring the strain with strain gauges (Niemczura & Ravi-Chandar, 2006). Nevertheless, there is a lack of experimental data on the measurements of the evolution of the phase transformation fronts at impact together with the stress-strain state.
The experimental set-up presented in Fig. 4 enables to obtain temperature measurements along the sample at different moments when it is deformed at impact strain rates, on the order of 10^{1}-10^{2} s^{-1}, while it is known the stress-strain state. The observation of the phase transformation fronts at impact could help in providing a better understanding of the stress induced martensitic transformation at high strain rates. Moreover, the use of the infrared thermographic technology simultaneously with conventional testing machines, allows the observation of fronts in the range of low strain rates, 10^{-4}-10^{-1} s^{-1}, which extends the understanding of the martensitic transformation as a function of the strain rate.

5.1 Phase transformation fronts evolution at impact strain rates

First of all it has been observed the evolution of the phase transformation fronts at impact for a complete martensitic transformation restricting the maximum strain achieved during deformation in order to avoid the elasto-plastic deformation of the martensitic phase, which may cause defects in the crystal lattice. For this, thermographic pictures were taken each 0.8 ms along a NiTi wire deformed up to 6.5% at a strain rate of 10 s^{-1}, Fig. 15(a). While the austenitic phase remains at nearby ambient temperature, the martensitic phase temperature rises due to the exothermic character of the forward transformation. During the unloading, the endothermic character of the reverse transformation cools again the austenitic phase to room temperature while the martensitic phase remains at higher temperature since the high speed of the impact event makes unable the releasing of the transformation heat from this phase. Then, the temperature difference between the austenite phase (low temperature) and martensite phase (high temperature) makes clearly visible the phase transformation fronts and their evolution in Fig. 15.

After the elastic deformation of the austenitic phase, the nucleation of the martensitic phase takes place at both ends of the sample near the grips, where stress concentrations are unavoidable. As the strain is increased, the two transformation fronts progress along the sample until completing the transformation. During unloading, the reverse transformation takes place at points where the forward transformation was finished. The confluence of the forward transformation fronts originates a discontinuity in the crystal lattice which is favourable for the nucleation of reverse transformation.

At impact no more nucleations were observed so that only the phase transformation fronts arising from the mentioned nucleations appear. This feature shows that the martensitic transformation at impact is inhomogeneous, similarly to that observed at very low strain rates, on the order of 10^{-4} s^{-1}, when the deformation may be considered as an isothermal process (Shaw & Kyriakides, 1995), going against the trend shown in some works (Shaw & Kyriakides, 1997; Chang et al., 2006), where the phase transformation fronts are multiplied as the strain rate is increased in the range 10^{-4} - 10^{-2} s^{-1}. This change of trend will be discussed in the next section.

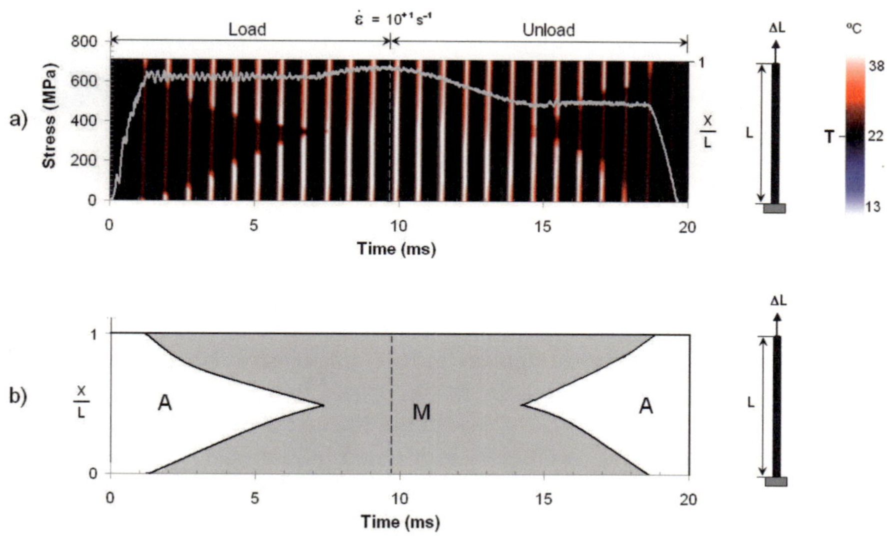

Fig. 15. a) Simultaneous evolution of the phase transformation fronts and the stress during an impact test (ε_{max}=6.5%), b) schematic phase diagram A(austenite)-M (martensite).

With the aim to study in depth the progress of the transformation, the temperature profiles measured along the central axis of the wire for each thermographic picture of the Fig. 15(a) were measured. Each of these profiles, represented by black lines in Fig. 16(b and c), correspond to a specific stress-strain state during the transformation Fig. 16(a), and they form a three-dimensional image that represents the gradient of temperatures between phases and the evolution of transformation fronts.

The short time during deformation at impact, on the order of a few milliseconds, makes unable to reach a steady state in the distribution of the temperature along the sample. The deformation time is so reduced that the heat transfer is confined to a small area between phases while the already transformed region remains at a higher temperature because the heat cannot be released to the surroundings or to other zones of the sample. This leads to a transient regime which shows a temperature gradient between the phases.

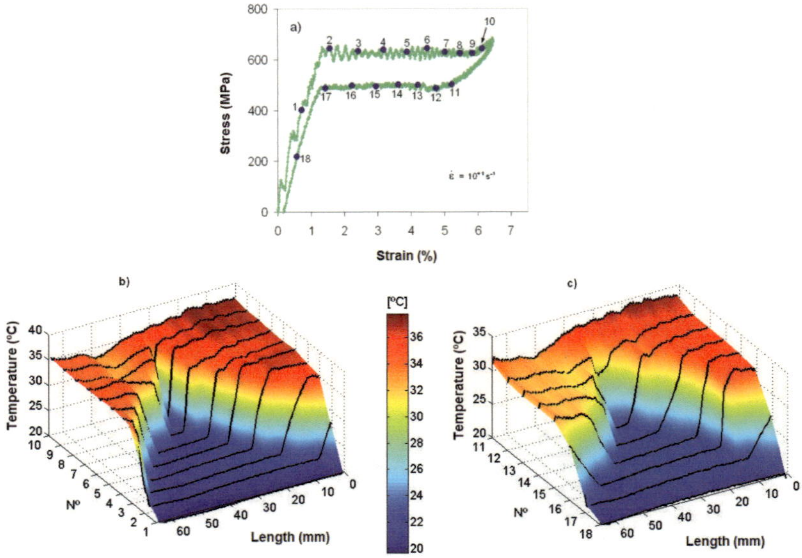

Fig. 16. Evolution of the complete stress induced martensitic transformation at impact strain rates, a) stress-strain diagram, b) temperature evolution during the forward transformation, b) temperature evolution during the reverse transformation.

When the deformation is high enough to induce the elasto-plastic deformation of the martensitic phase, some defects may arise in the crystal lattice modifying the evolution of the reverse transformation. The small stress fields formed around defects during loading may retain certain amount of preferential oriented martensite that lowers the stress necessary to perform the reverse transformation during the unloading and assisting the generation of new nucleations. This effect is shown in Fig. 17, that besides the first step of the reverse transformation (t=16 ms), another retransformed region appears later (for X/L close to 1 and t=17.5 ms). This is also clearly visible in the temperature map of the Fig. 18(c), in which a new temperature gradient appears at point 11 for longitude x=0.

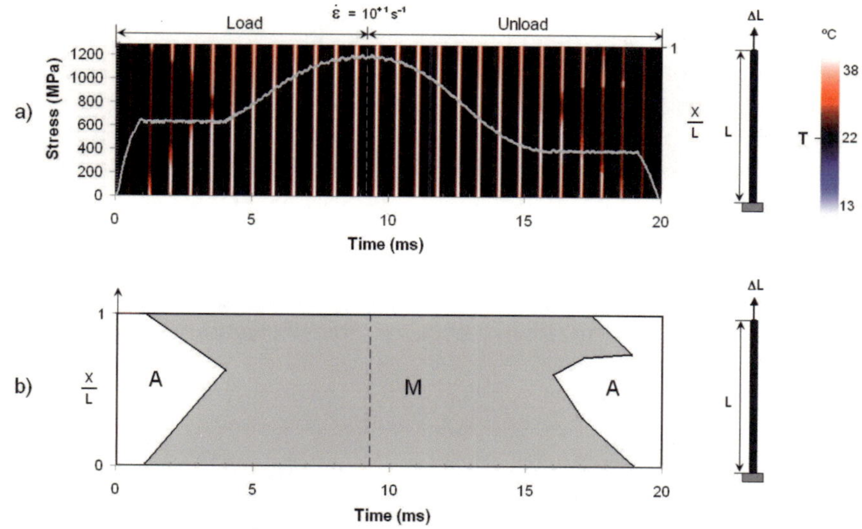

Fig. 17. a) Simultaneous evolution of the phase transformation fronts and the stress during an impact test (ε_{max}=10%), b) schematic phase diagram A(austenite)-M (martensite).

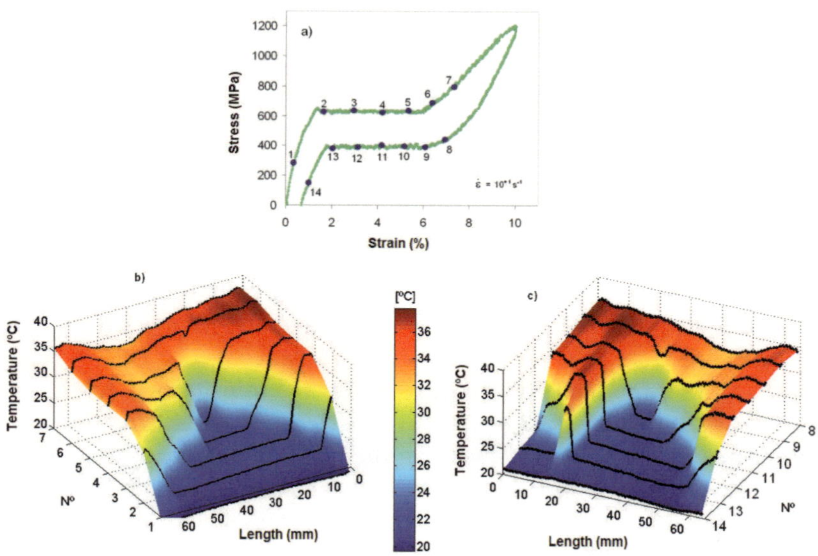

Fig. 18. Evolution of the elasto-plastic deformation of the stress induced martensitic phase at impact strain rates, a) stress-strain diagram, b) temperature evolution during the forward transformation, b) temperature evolution during the reverse transformation.

5.1 Phase transformation fronts evolution as function of the strain rate

In this section the homogeneity of the B2-B19′ martensitic transformation is discussed as function of the strain rate. This quality may be evaluated in terms of the number of the B19′ phase nucleations.

It is well known that this transformation is inhomogeneous at strain rates lower than 10^{-4} s^{-1} (Shaw & Kyriakides, 1995). The trend observed in the literature shows that the number of phase transformation fronts is multiplied as the strain rate is increase within the range from 10^{-4} s^{-1} to 10^{-2} s^{-1} (Shaw & Kyriakides, 1995; Chang et al., 2006), going against the results obtained in the previous section at impact strain rates, 10^{+1} s^{-1}, where the transformation appears to be inhomogeneous, without multiple transformation fronts. In order to study in depth this evolution, the number of nucleations was observed at intermediate strain rates.
In fact, when the strain rate is low enough, only phase transformation fronts nucleated at grips are observed, Fig. 19. Once the nucleation takes place, the stress necessary to its propagation is lower than that necessary to originate a new nucleation, so that the initial front propagates leading to an inhomogeneous transformation as is corroborated in literature (Shaw & Kyriakides, 1995).

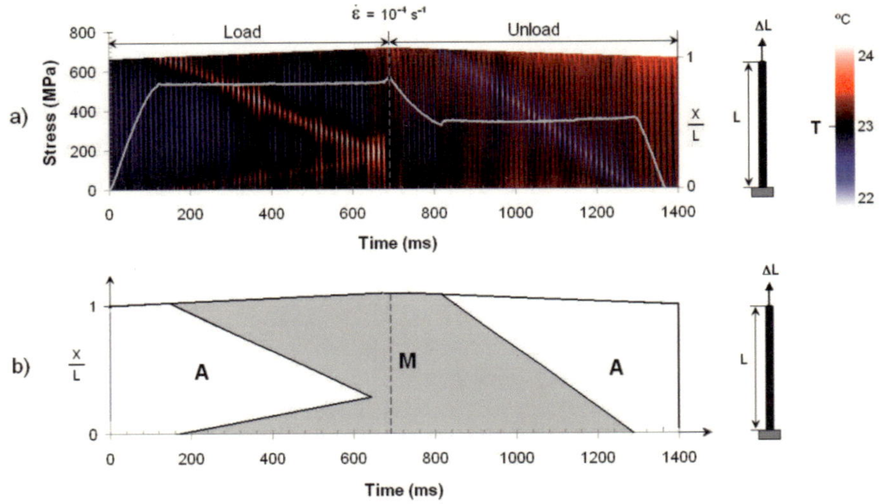

Fig. 19. a) Simultaneous evolution of the phase transformation fronts and the stress at a strain rate of 10^{-4} s^{-1}, b) schematic phase diagram A(austenite)-M (martensite).

At higher strain rates, the temperature increases in the transformed zones, since the time available to release the transformation heat to the sourrandings is reduced. This local rise of temperature increases the stress necessary for the propagation of the active front due to the inherent sensibility of this transformation with the temperature. When the strain rate is high enough, the temperature may reach such a high value that the propagation stress of the active front is higher than that necessary to allow another nucleation in a cooler region of the sample far away of the active front, so another nucleations may arise, Fig. 20.

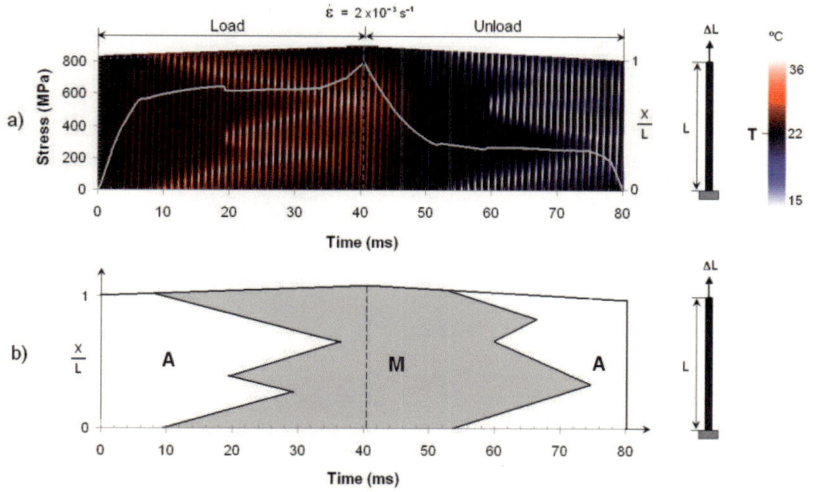

Fig. 20. a) Simultaneous evolution of the phase transformation fronts and the stress at a strain rate of $2x10^{-3}$ s^{-1}, b) schematic phase diagram A(austenite)-M (martensite).

As the strain rate increases, this effect is accentuated, and additional nucleations appear, Fig. 21. Moreover, just at the moment at which the new nucleations appear, a small drop in stress is shown, similarly to that observed in other works (Leo et al., 1993, Shaw & Kyriakides, 1997, Chang et al., 2006). This is due to the sudden and localized small increase of strain which occurs at the nucleation sites. On the other hand, the multiplication of the nucleations implies a reduction of the fronts speed in order to maintain the consistency of the global strain rate, resulting in a lower self-heating. This reduces the immediate possibility of the emergence of a new nucleation, which is postponed until the material is reheated.

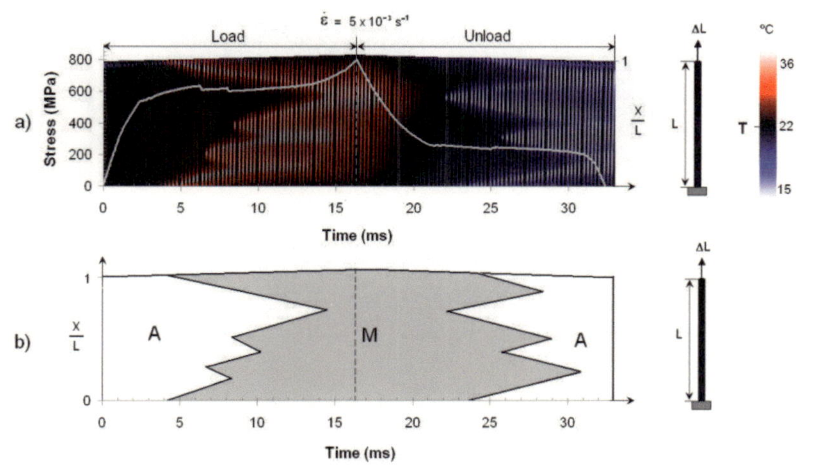

Fig. 21. a) Simultaneous evolution of the phase transformation fronts and the stress at an strain rate of $5x10^{-3}$ s^{-1}, b) schematic phase diagram A(austenite)-M (martensite).

As the strain rate increases, the time available for releasing the transformation heat to the surroundings is more and more reduced. This enhances the effect of the self-heating, and the number of nucleations continues multiplying. In this case, the temperature at the beginning of the reverse transformation may be even greater than the initial temperature of the specimen, Fig. 22.

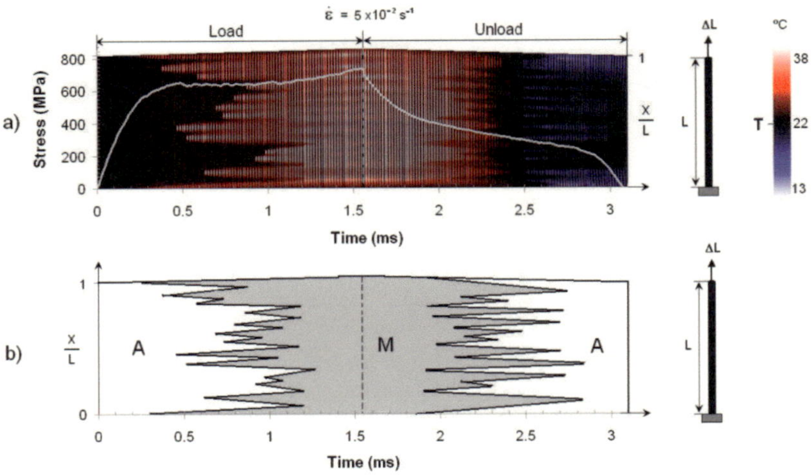

Fig. 22. a) Simultaneous evolution of the phase transformation fronts and the stress at an strain rate of $5x10^{-2}$ s^{-1}, b) schematic phase diagram A(austenite)-M (martensite).

Nevertheless, when the strain rate is high enough, 10^{-1} s^{-1}, the new nucleations appear near the active fronts, as is shown in Fig. 23. This may mean that at impact, Fig. 15 and Fig. 17, multiple transformations also occur, but so near the active fronts that they may are visible as a single one.

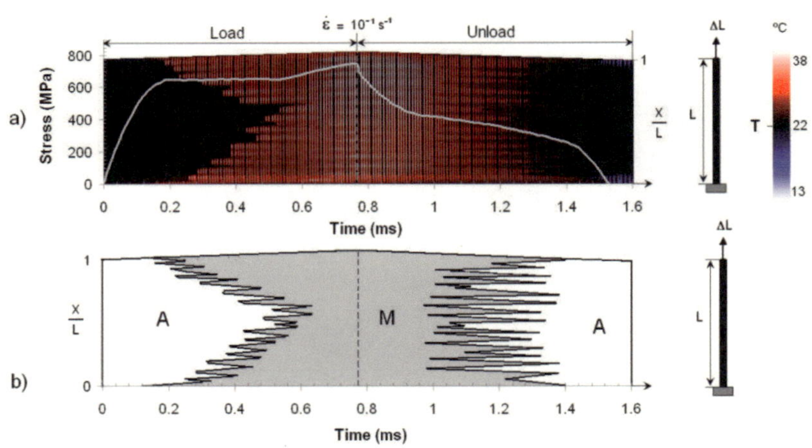

Fig. 23. a) Simultaneous evolution of the phase transformation fronts and the stress at a strain rate of 10^{-1} s^{-1}, b) schematic phase diagram A(austenite)-M (martensite).

6. References

Adharapurapu, R.R.; Jiang, F.; Vecchio, K.S. & Gray III, G.T. (2006). Response of NiTi shape memory alloy at high strain rate: A systematic investigation of temperature effects on tension-compression asymmetry. Acta Materialia, v 54, n 17, p 4609–4620.

Auricchio, F.; Faravelli, L.; Magonette, G. & Torra, V. (2001). Shape Memory Alloys. Advances in Modelling and Applications, CIMNE, Barcelona.

Boyce, B.L. & Crenshaw, T.B. (2005). Servohydraulic Methods for Mechanical Testing in the Sub-Hopkinson Rate Regime up to Strain Rates of 500 1/s. SAND2005-5678, Sandia National Laboratories.

Bragov, A.M.; Lomunov, A.K. & Sergeichev, I.V. (2006). High-speed behavior of some shape memory alloys. AIP Conference Proceedings, v 845, n 1, p 705–708.

Chang, B.C.; Shaw, J.A. & Iadicola, M.A. (2006). Thermodynamics of shape memory alloy wire: Modeling, experiments, and application. Continuum Mechanics and Thermodynamics, v 18, n 1–2, p 83–118.

Chen, W.W.; Wu, Q.; Kang, J.H. & Winfree, N.A. (2001). Compressive superelastic behavior of a NiTi shape memory alloy at strain rates of 0.001–750 s–1. International Journal of Solids and Structures, v 38, n 50–51, p 8989–8998.

Chrysochoos, A.; Lobel, M. & Maisonneuve, O. (1995). Thermo-mechanical coupling of pseudoelastic behaviour of CuZnAl and NiTi alloys. Comptes Rendus de l'Academie des Sciences, Serie II (Mecanique–Physique–Chimie–Astronomie), v 320, n 5, p 217–223.

Dolce, M. & Cardone, D. (2001). Mechanical behaviour of shape memory alloys for seismic applications 2. Austenite NiTi wires subjected to tension. International Journal of Mechanical Sciences, v 43, n 11, p 2657–2677.

Duerig, T.W.; Melton, K.N.; Stöckel, D. & Wayman, C.M. (1990). Engineering aspects of shape memory alloys, Butterworth- Heinemann, London.

Entermeyer, D.; Patoor, E.; Eberhardt, A. & Berveiller, M. (2000). Strain rate sensitivity in superelasticity. International Journal of Plasticity, v 16, n 10–11, p 1269–1288.

Funakubo, H. (1987). Application of shape memory alloys, In: Shape Memory Alloys, Funakubo, H. & Kennedy J.B. (Eds.), second ed., pp. 227–269, Gordon and Breach Science, New York.

Iadicola, M.A. & Shaw, J.A. (2007). An experimental method to measure initiation events during unstable stress-induced martensitic transformation in a shape memory alloy wire. Smart Materials and Structures, v 16, n 1, p 155–169.

Lagoudas, D.C., (2008). Shape Memory Alloys: Modeling and Engineering Applications, Spriger, University College Station TX, USA.

Leo, P.H.; Shield, T.W. & Bruno, O.P. (1993). Transient heat transfer effects on the pseudoelastic behavior of shape-memory wires. Acta Metallurgica et Materialia, v 41, n 8, p 2477-2485.

Li, H., Mao, C.; Ou, J. (2005). Strain self-sensing property and strain rate dependent constitutive model of austenitic shape memory alloy: Experiment and theory. Journal of Materials in Civil Engineering, v 17, n 6, p 676–685.

Lin, P.H.; Tobushi, H.; Tanaka, K.; Hattori, T. & Ikai, A. (1996). Influence of strain rate on deformation properties of TiNi shape memory alloy. JSME International Journal, Series A, v 39, n 1, p 117–123.

Liu, Y., Li, Y., Xie, Z. & Ramesh, K.T. (2002). Dynamic deformation of shape- memory alloys: evidence of domino detwinning? Philosophical Magazine Letters, v 82, n 9, p 511-517.

Miller, D.A.; Thissell, W.R. & Macdougall, D.A.S. (2000). Dynamic tensile plasticity and damage evolution in shape-memory Ni-Ti. Journal de Physique IV, v 10, n 9, p 341-346.

Niemczura, J. & Ravi-Chandar, K. (2006). Dynamics of propagating phase boundaries in NiTi. Journal of the Mechanics and Physics of Solids, v 54, n 10, p 2136-2161.

Otsuka, K. & Wayman, C.M. (1998). Shape Memory Materials. Cambridge Unversity Press, Cambridge.

Pieczyska, E.A.; Tobushi, H.; Gadaj, S.P. & Nowacki, W.K. (2006a). Superelastic deformation behaviors based on phase transformation bands in TiNi shape memory alloy. Materials Transactions, v 47, n 3, p 670-676.

Pieczyska, E.A.; Gadaj, V.; Nowacki, W.K. & Tobushi, H. (2006b). Phase-transformation fronts evolution for stress- and strain-controlled tension tests in TiNi shape memory alloy. Experimental Mechanics, v 46, n4, 531-542.

Pieczyska, E.; Nowacki, W.; Sakuragi, T. & Tobushi, H. (2007). Superelastic deformation properties of TiNi shape memory alloy. Key Engineering Materials, 340-341, pt.2, p 1211-1216.

Prahlad, H. & Chopra, I. (2000). Experimental characterization of Ni-Ti shape memory alloy wires under uniaxial loading conditions. Journal of Intelligent Material Systems and Structures, v 11, n 4, p 272-282.

Qiu, Z.-X.; Yao, X.T.; Yuan, J. & Soutis, C. (2006). Experimental research on strain monitoring in composite plates using embedded SMA wires. Smart Materials and Structures, v 15, n 4, p 1047-1053.

Schmidt, I. (2006). A phenomenological model for superelastic NiTi wires based on plasticity with focus on strain-rate dependency caused by temperature. Transactions of the ASME. Journal of Engineering Materials and Technology, v 128, n 3, p 279-284.

Shaw, J.A. & Kyriakides, S. (1995). Thermo-mechanical aspects of NiTi. Journal of the Mechanics and Physics of Solids, v 43, n 8, p 1243-1281.

Shaw, J.A. & Kyriakides, S. (1997). On the nucleation and propagation of phase transformation fronts in a NiTi alloy. Acta Materialia, v 45, n 2, p 683-700.

Tobushi, H.; Shimeno, Y.; Hachisuka, T. & Tanaka, K. (1998). Influence of strain rate on superelastic properties of TiNi shape memory alloy. Mechanics of Materials, v 30, n 2, p 141-150.

Tsoi, K.A., Stalmans, R., Schrooten, J., Wevers, M. & Mai, Y. (2003). Impact damage behaviour of shape memory alloy composites. Materials Science & Engineering A, v A342, n 1-2, p 207-215.

Vitiello, A.; Giorleo, G. & Morace, R.E. (2005). Analysis of thermo-mechanical behaviour of Nitinol wires with high strain rates. Smart Materials and Structures, v 14, n 1, p 215-221.

Wu, K,; Yang, F.; Pu, Z. & Shi, J. (1996). Effect of strain rate on detwinning and superelastic behavior of NiTi shape memory alloys. Journal of Intelligent Material Systems and Structures, v 7, n 2, p 138-144.

Xu, R.; Cui, L. & Zheng, Y. (2006). The dynamic impact behavior of NiTi alloy. Materials Science and Engineering A, v 438-440, n SPEC. ISS., p 571-574.

Zurbitu, J.; Castillo, G.; Urrutibeascoa, I. & Aurrekoetxea, J. (2009a). Low-energy tensile-impact behavior of superelastic NiTi shape memory alloy wires. Mechanics of Materials, v 41, n 9, p 1050-1058.

Zurbitu, J.; Castillo, G.; Aretxabaleta, L. & Aurrekoetxea, J. (2009b). Phase transformation fronts propagation during the Stress Induced Martensitic transformation in NiTi Shape Memory Alloy wires at impact strain rates. ESOMAT 2009 - The 8th European Symposium on Martensitic Transformations, [06038], edited by P. Šittner, L. Heller and V. Paidar, published by EDP Sciences.

Zurbitu, J.; Kustov, S.; Castillo, G.; Aretxabaleta, L.; Cesari, E. & Aurrekoetxea, J. (2009c). Instrumented tensile-impact test method for shape memory alloy wires. Materials Science and Engineering A, v 524, n 1-2, p 108-111.

3

Bending Deformation and Fatigue Properties of Precision-Casting TiNi Shape Memory Alloy Brain Spatula

Hisaaki Tobushi[1], Kazuhiro Kitamura[2],
Yukiharu Yoshimi[3] and Kousuke Date[1]
1Department of Mechanical Engineering, Aichi Institute of Technology
1247 Yachigusa, Yakusa-cho, Toyota, 470-0392, Japan
[2]Department of Technology Education, Aich University of Education,
Hirosawa 1, Igaya-cho, Kariya, 448-8542, Japan
[3]Yoshimi, Inc., 1-43 Kitasaki-cho, Obu, 474-0002, Japan
E-mail: tobushi@aitech.ac.jp

Abstract

In order to develop a brain spatula or a brain retractor made of a shape memory alloy (SMA), the bending characteristics of the brain spatula of TiNi SMA made by the precision casting were discussed based on the tensile deformation properties of the existing copper and the TiNi rolled-SMA. The fatigue properties of these materials were also investigated by the pulsating- and alternating-plane bending fatigue tests. The results obtained can be summarized as follows. (1) Based on the yield stress and the modulus of elasticity of the copper and the SMA, the bending deformation properties of an SMA-brain spatula were estimated by assuming the condition to use the brain spatula as the bending of a cantilever. With respect to the SMA-brain spatula for the same length and width as the existing copper-brain spatula, if the thickness of the new cast-SMA brain spatula is 1.2 times and that of the conventional rolled-SMA brain spatula is 1.3 times as large as that of the existing copper one, the SMA-brain spatula can hold the same bending rigidity and can be bent by almost the same force as the existing copper one. (2) With respect to the alternating- and pulsating-plane bending fatigue, the fatigue life of both the copper and the SMAs in the region of low-cycle fatigue is expressed by a power function of the maximum bending strain. The fatigue life of the conventional rolled SMA and the new cast SMA is longer than that of the existing copper. The fatigue life of the new cast and rolled SMAs in the pulsating-plane bending is longer than that in the alternating-plane bending. (3) The above mentioned characteristics of the SMA-brain spatula obtained in this study will be substantially applied to the development not only for the brain spatula but also for other retractors and instruments used in other surgery operations.

Keywords: Shape Memory Alloy, Brain Spatula, Titanium-Nickel Alloy, Fatigue, Precision Casting, Bending

1. Introduction

The shape memory effect (SME) and superelasticity (SE) appear in a shape memory alloy (SMA). The strain of several percents is recovered by heating for the SME and by unloading for the SE. These characteristics occur due to the martensitic transformation (MT) and its reverse transformation. Large recovery stress, energy dissipation and storage which appear owing to the MT can be used in the SMA [1-5]. The development of applications of the SMA as the intelligent materials has therefore attracted the worldwide attention. The SMA has been used in the wide fields of industry, electric products, medical devices, leisure and so on.

A brain spatula or a brain retractor is used as an instrument in a surgical operation of a brain. This instrument is used to hold the opened state of the brain during the operation of the cerebral tumor which is located in the inner part of the brain. An image of the brain spatula used in the operation is shown in figure 1. As shown in figure 1, the brain spatula is used in the bent form, fitting it to the shape and depth of each patient's brain. After the operation, the brain spatula is struck with a mallet to recover the original flat plane followed by the treatment in a sterilizer by heating and is used again thereafter. The main material used to the existing brain spatula is copper. Since the irrecoverable unevenness appears on the surface of the brain spatula after the use, it is disposed after using several times. If the SMA is used for the brain spatula, the brain spatula used in the bent form regains its original flat shape automatically based on the SME through the treatment of sterilization by heating in the autoclave. Therefore, since the SMA-brain spatula not only saves the time to strike with a mallet to regain the original flat shape but also recovers its original shape automatically, the appearance of the uneven plastic deformation is inhibited and the spatula can be used times out of number.

The TiNi SMA products made by the precision casting utilizing the lost-wax process from a self-combustion high-temperature synthesis method have been newly developed [6]. The brain spatula needs various shapes corresponding to the brain of each patient. TiNi SMA is very hard to form the complex shape of the products. An example of the TiNi SMA-brain spatula made by the precision casting is shown in figure 2. The TiNi SMA-brain spatula with complex shape can be easily produced by the precision casting.

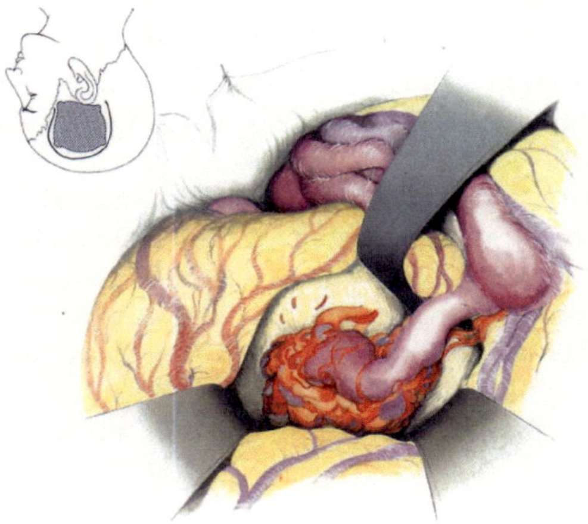

Fig. 1. Image of brain spatula in operation

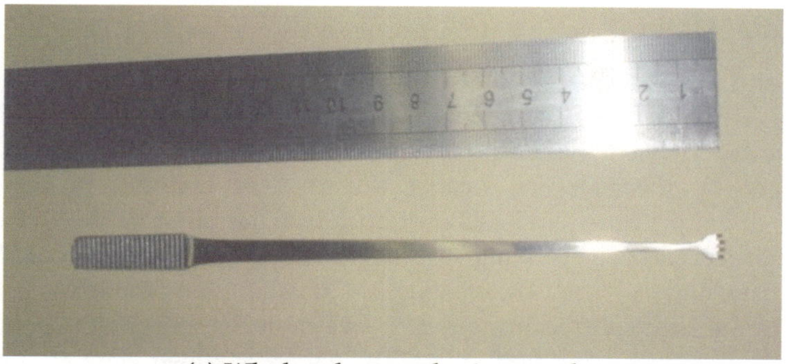

(a) Whole rake-type brain spatula

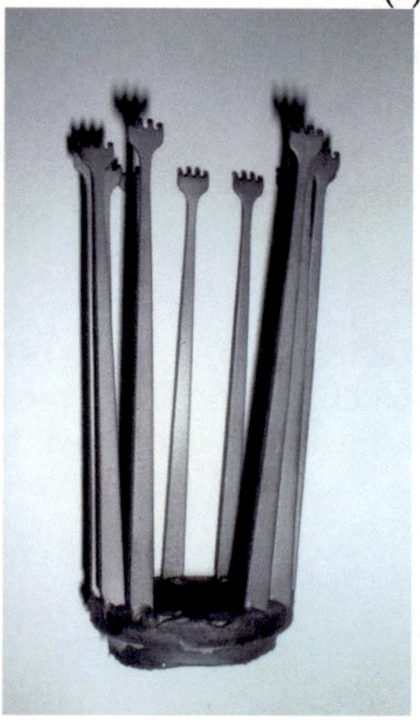

(b) Brain spatula just after precision casting

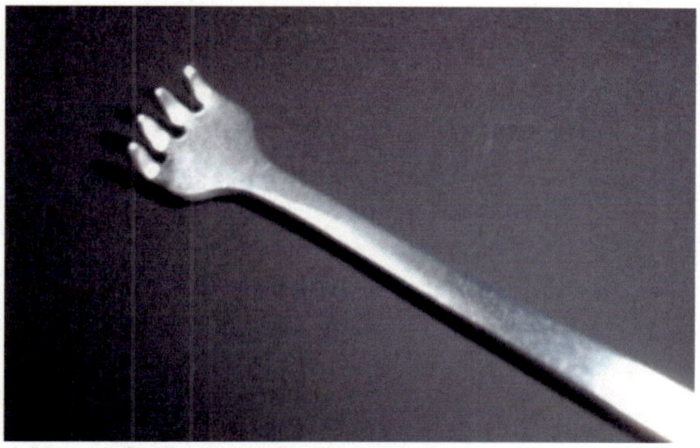

(c) Tip of rake-type brain spatula

Fig. 2. Example of TiNi SMA-brain spatula made by the precision casting

The required conditions for the brain spatula are the fact that it can be bent in an arbitrary shape to fit it to the brain of a patient in the operation and it has enough rigidity to hold the opened state of the brain during the operation. These characteristics can be prescribed based on the bending deformation properties of the brain spatula including the bending rigidity. In order to evaluate the reliability of the brain spatula from the view point of safety, the fatigue properties of the materials are crucial. The brain spatula is subjected to cyclic plane bending. The plane-bending fatigue properties of TiNi SMA for the brain spatula have not been reported till now.

In the present study, in order to develop the SMA-brain spatula, the stress-strain relations of the new cast TiNi SMA, the conventional rolled TiNi SMA and the copper used for the existing brain spatula were examined by the tension tests. The shape and dimension which are required for the SMA-brain spatula were investigated based on the bending deformation properties of the strip cantilever. The fatigue properties of the materials which are very important in the practical cyclic use were also investigated.

2. Experimental Method

2.1 Materials and specimens

The materials used in the experiment were the new cast Ti-49.7at%Ni SMA, the conventional rolled Ti-50.0at%Ni SMA, and the existing copper brain spatula. The new cast SMA was made by the precision casting utilizing the lost-wax process from a self-combustion high-temperature synthesis method [6]. A flat plane of the rolled and the cast SMA was shape-memorized by fixing in a flat plane for 40 min at 753 K in the furnace followed by quenching in water. The starting and finishing temperatures of the MT M_s and M_f and those of the reverse transformation A_s and A_f of the SMAs were obtained from the DSC (differential scanning calorimetric) test. The values obtained were M_s = 326 K, M_f= 312 K, A_s = 342 K, A_f= 365 K for the rolled SMA and M_s = 358 K, M_f= 283 K, A_s = 314 K, A_f= 386 K for the cast SMA. The specimens used in the tension test were the uniform rectangular bars with a thickness of t = 1.0 mm, a width of w = 1.2 mm and a length of l = 160 mm for the rolled and cast SMAs, and t = 1.0 mm, w = 8.5 mm and l=140 mm for the copper. In the fatigue test, all specimens were the bars with t = 1.0 mm, w=1.2 mm and l = 80 mm.

2.2 Experimental apparatus

An SMA characteristic testing machine was used for the tension test [7]. The testing machine was composed of a tension machine for loading and unloading and a heating-cooling device to control temperature. Displacement of the specimen was measured by an extensometer with a gauge length of 50 mm.

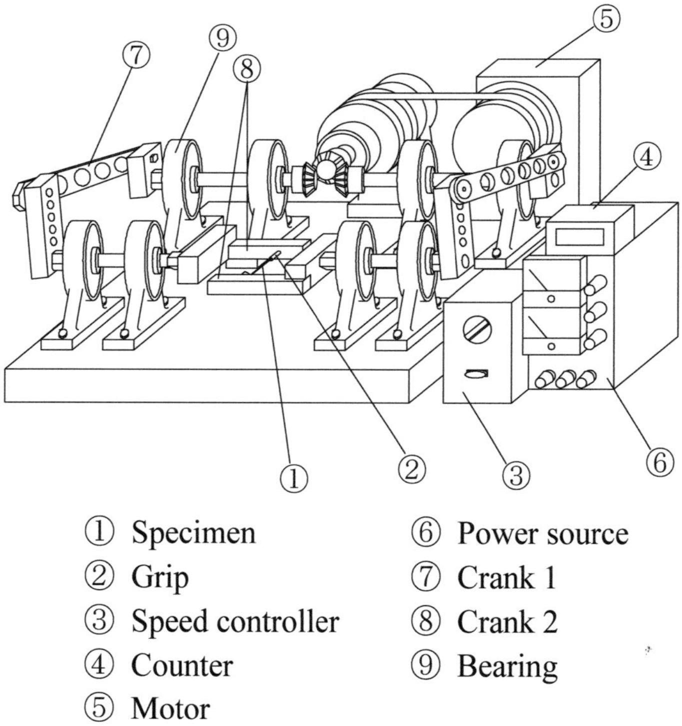

Fig. 3. Experimental apparatus for alternating-plane bending fatigue test

An alternating-plane bending fatigue testing machine [8] used in the fatigue test is shown in figure 3. In the fatigue testing machine, the maximum bending strain occurs at the central part of the specimen □, and the tensile and compressive bending strains occur alternately. A pulsating-plane bending fatigue testing machine [9] was also used in the fatigue test. In the alternating- and pulsating-plane bending tests, the maximum bending strain on the surface of the specimen was selected and the number of cycles to failure was obtained under a constant frequency. A scanning electron microscope (SEM) was used to observe the fracture surface of specimen.

2.3 Experimental procedure

The tension tests were carried out under a constant strain rate in air at room temperature below the M_f point of the SMA bars. Since the yielding of the SMA occurs under low stress in the case of the M-phase, the SMA-brain spatula can be easily bent by small force. For the M-phase SMA bar, the residual strain appears after unloading. The SMA bars with the residual strain were heated up to temperatures above the A_f point under no load. In the heating process of the SMA bars, the residual strain diminishes due to the reverse transformation between the A_s and A_f points.

The alternating- and pulsating-plane bending fatigue tests were carried out in air at room temperature. Every specimen was fractured in the central part of its length between grips where the bending strain takes the maximum value. The maximum bending strain ε_m on the surface of the specimen was obtained from the radius of curvature at the point of fracture. The frequency f was 8.33 Hz (500 cpm) and 3.33 Hz (200 cpm).

3. Deformation properties of materials used for brain spatula

3.1 Tensile deformation properties

The stress-strain curves of the copper, the rolled and the cast SMAs obtained from the tension test under a strain rate of $d\varepsilon/dt = 2\times10^{-4}$ s^{-1} are shown in figure 4. As can be seen in figure 4, the linear elastic deformation occurs in the initial loading stage and the yielding occurs thereafter. The modulus of elasticity E determined from a slope of the initial stress-strain curves is E = 40 GPa for the rolled SMA, E = 54 GPa for the cast SMA and E = 95 GPa for the copper. Approximating the elastic and yield regions of the stress-strain curves by two straight lines, the yield stress σ_y was determined from an intersection of two lines. The yield stress was σ_y = 68 MPa for the rolled SMA, σ_y = 168 MPa for the cast SMA and σ_y = 240 MPa for the copper. In the case of the copper, the deformation above a strain of 0.2 % occurs due to the plastic deformation with dislocations. Therefore, in the unloading process from a strain of 4 %, strain is recovered by 0.25 % due to the elastic deformation and the residual strain appears as the permanent strain. In the case of the SMAs, since the material is in the M-phase at room temperature below the M_f point, the yielding occurs due to the rearrangement of the M-phase. In the unloading process from a strain of 4 %, strain is recovered by 0.6 % and 0.8 % for the rolled and the cast SMAs, respectively, and the residual strain of 3.4 % and 3.2 % appears after unloading.

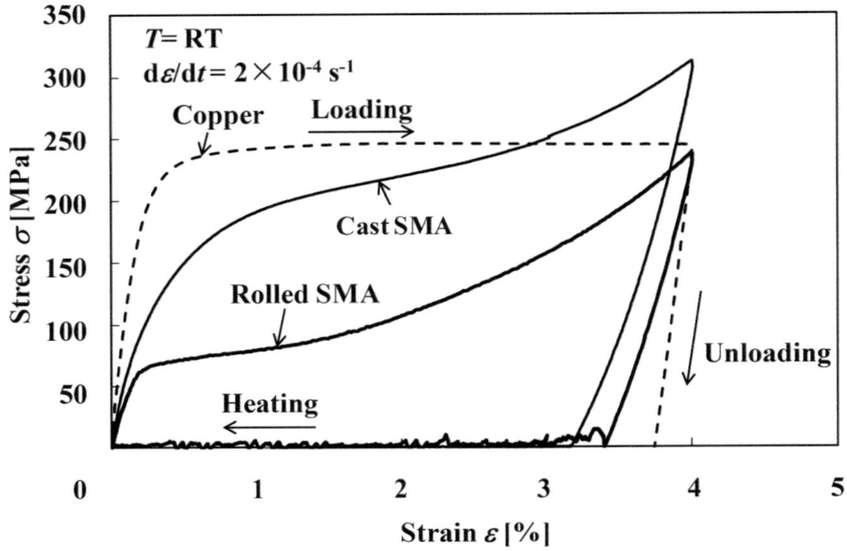

Fig. 4. Stress-strain curves of copper, rolled SMA and cast SMA used for brain spatula in tension

The relationship between strain and temperature of the cast and rolled SMAs obtained from the tension test with unloading followed by heating under no load is shown in figure 5. The symbols $A_{s:C}$, $A_{s:R}$, $A_{f:C}$ and $A_{f:R}$ shown in figure 5 represent the reverse-transformation starting and finishing temperatures of the cast and rolled SMAs, respectively. In the heating process under no load, strain starts to be recovered gradually around A_s and disappears perfectly around A_f. The SME to show this strain recovery occurs due to the reverse transformation from the M-phase to the parent (austenite) phase.

3.2 Comparison of characteristic values for deformation

The values of the modulus of elasticity E, the yield stress σ_y and the yield strain ε_y of the copper, the rolled and the cast SMAs obtained from the tension test are shown in table 1. As can be seen from table 1, both the modulus of elasticity and the yield stress of the rolled and the cast SMAs are lower than those of the copper. These differences affect the deformation residence to bend the brain spatula and the bending rigidity to hold the opened shape of the brain during the operation. In the next section, the bending deformation properties of the brain spatulas made of these materials will be discussed.

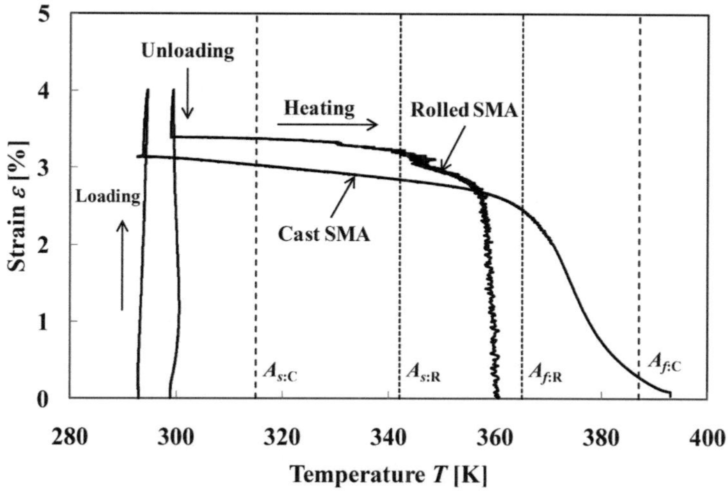

Fig. 5. Strain-temperature curves of rolled SMA and cast SMA in tension with unloading followed by heating under no-load

Table 1. Values of modulus of elasticity, yield stress and yield strain for copper, rolled SMA and cast SMA

	Copper	Rolled SMA (M-phase)	Cast SMA (M-phase)
Modulus of elasticity E [GPa]	95	40	54
Yield stress σ_y [MPa]	240	68	168
Yield strain ε_y [%]	0.25	0.17	0.34

4. Bending characteristics of copper- and SMA-brain spatulas

4.1 Elastic bending property of brain spatula

In order to design the SMA-brain spatula, it is important to evaluate the force which is necessary for a medical doctor to bend the brain spatula and the bending rigidity which is required to hold the opened shape of the brain during the operation. Discussing these bending deformation properties of the copper- and the SMA-brain spatulas, the characteristics required for the SMA-brain spatula will be clarified. Let us treat the bending of the existing brain spatula as the bending of a cantilever made of the strip with a uniform rectangular cross-section. The length of the strip is expressed by l, the width of the cross-section by w and the thickness by t. In order to hold the displacement of the opened part of the brain during the operation, it is required for the brain spatula to keep the bent form (see figure 1). This requirement can be evaluated by the maximum deflection of the cantilever made of these materials. Let us consider the condition to obtain the same maximum deflection y_{max} by the same force F applied at the top of the cantilever. The maximum deflection of the cantilever y_{max} can be expressed by using the second moment of area $I_z = wt^3/12$ from the theory of elasticity as follows

$$y_{\max} = \frac{Fl^3}{3EI_z} = \frac{4Fl^3}{Ewt^3} \tag{1}$$

If the maximum deflection y_{max} expressed by Eq. (1) for the copper and the SMA strips subjected to the same force F coincides, the following equation is obtained

$$y_{\max} = \frac{4Fl_{Cu}^3}{E_{Cu} \cdot w_{Cu} \cdot t_{Cu}^3} = \frac{4Fl_{SMA}^3}{E_{SMA} \cdot w_{SMA} \cdot t_{SMA}^3} \tag{2}$$

From the view point of the practical operation of the brain, the width w and the length l of the SMA-brain spatula are expected to take the same values as those of the existing copper-brain spatula. Therefore, let us consider the condition that the length l and the width w of both spatulas coincide and only the thickness t differs. The thickness of SMA t_{SMA} becomes from Eq. (2) as follows

$$t_{SMA} = \sqrt[3]{\frac{E_{Cu}}{E_{SMA}}} t_{Cu} \tag{3}$$

The values of the modulus of elasticity E of the copper and the SMAs shown in table 1 are substituted in Eq. (3). As a result, if the thickness of the rolled-SMA spatula is 1.3 times and that of the cast-SMA spatula is 1.2 times as large as that of the copper one, the bending rigidity of both copper and SMA spatulas coincides and the same deflection can be obtained by the SMA-brain spatulas.

4.2 Plastic deformation resistance for bending

In the operation of the brain, a medical doctor bends the brain spatula to fit it to the shape and depth of the opened part of each patient's brain. The force to bend the brain spatula is valuated by the force applied at the top of the cantilever to obtain the required maximum bending strain. The yield region appears in the surface element of the cantilever during bending. The elastic and yield regions of the strip in bending are shown in figure 6. The schematic distributions of the bending strain and stress in the strip are shown in figures 7 and 8, respectively. In these figures, the maximum bending strain, the yield strain and the yield stress are denoted by ε_m, ε_y and σ_y, respectively. It is assumed that the yield stress σ_y in tension is the same as that in compression and is constant in the yield region.

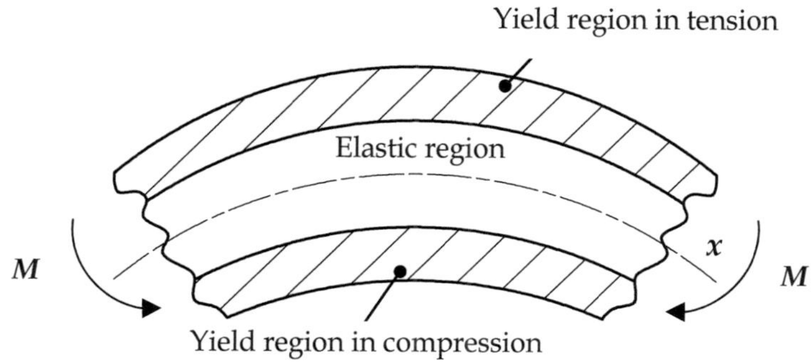

Fig. 6. Elastic and yield regions of strip in bending

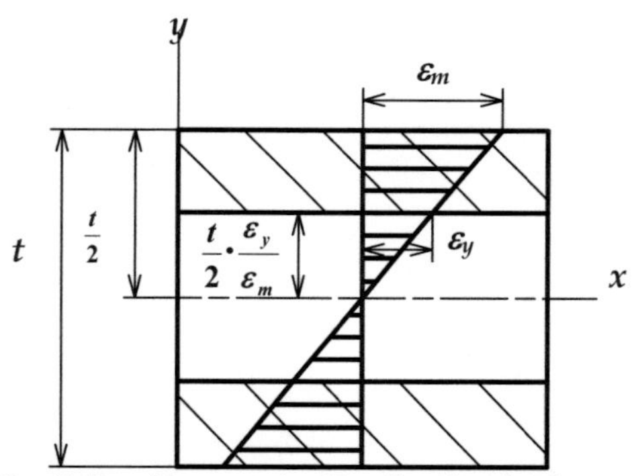

Fig. 7. Bending strain distribution in the strip

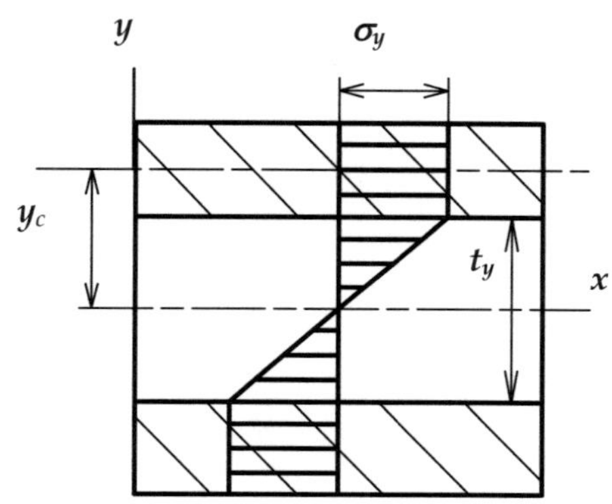

Fig. 8. Bending stress distribution in the strip

At first, let us consider the bending moment M_e which is required to obtain the elastic region of the strip in bending. The thickness of the elastic region t_y is $t_y = (\varepsilon_y/\varepsilon_m)t$. Since the maximum bending stress in the elastic region σ_{max} is the same as the yield stress σ_y, the bending moment required for the elastic region M_e is $M_e = \sigma_y \cdot Z$, where Z denotes the section modulus and $\boldsymbol{Z} = \boldsymbol{w}\boldsymbol{t}_{\boldsymbol{y}}^2/6$. Therefore, the bending moment M_e for the elastic region is

$$M_e = \frac{\sigma_y w t^2}{6}\left(\frac{\varepsilon_y}{\varepsilon_m}\right)^2 \tag{4}$$

Next, let us consider the bending moment M_y required to cause the yield region of the strip. The center of the yield region y_c from the neutral axis z is

$$y_c = \frac{t}{4}\left(1 + \frac{\varepsilon_y}{\varepsilon_m}\right) \tag{5}$$

The area A of the yield region in the tension side on the cross section is

$$A = \frac{wt}{2}\left(1 - \frac{\varepsilon_y}{\varepsilon_m}\right) \tag{6}$$

Considering the constant stress σ_y in the yield region and the symmetric condition in the tension and compression sides, the bending moment M_y required to cause the yield region is obtained as follows.

$$M_y = \sigma_y \cdot A \cdot 2y_c = \frac{\sigma_y w t^2}{4}\left\{1 - \left(\frac{\varepsilon_y}{\varepsilon_m}\right)^2\right\} \tag{7}$$

The total bending moment M required to bend the brain spatula is given by a sum of the bending moment M_e for the elastic region and the bending moment M_y for the yield region.

$$M = M_e + M_y = \frac{\sigma_y w t^2}{12}\left\{3 - \left(\frac{\varepsilon_y}{\varepsilon_m}\right)^2\right\} \tag{8}$$

Since the bending moment M is $\boldsymbol{M = Fl}$, the force required is

$$F = \frac{\sigma_y w t^2}{12l}\left\{3 - \left(\frac{\varepsilon_y}{\varepsilon_m}\right)^2\right\} \tag{9}$$

The condition, that the forces required to bend the copper-brain spatula and the SMA-brain spatula coincide, is as follows.

$$F = \frac{\sigma_{yCu} w_{Cu} t_{Cu}^2}{12 l_{Cu}}\left\{3 - \left(\frac{\varepsilon_{yCu}}{\varepsilon_{mCu}}\right)^2\right\} = \frac{\sigma_{ySMA} w_{SMA} t_{SMA}^2}{12 l_{SMA}}\left\{3 - \left(\frac{\varepsilon_{ySMA}}{\varepsilon_{mSMA}}\right)^2\right\} \tag{10}$$

If the length l, the width w and the maximum bending strain ε_m of both spatulas coincide and only the thickness differs, the thickness of the SMA-brain spatula is

$$t_{SMA} = t_{Cu}\left[\frac{\sigma_{yCu}\left\{3-\left(\frac{\varepsilon_{yCu}}{\varepsilon_m}\right)^2\right\}}{\sigma_{ySMA}\left\{3-\left(\frac{\varepsilon_{ySMA}}{\varepsilon_m}\right)^2\right\}}\right]^{\frac{1}{2}} \tag{11}$$

The yield stress σ_y and the yield strain ε_y of each material are shown in table 1. Using these values, the thickness of the SMA-brain spatula t_{SMA} was obtained from Eq. (11). The calculated results of t_{SMA} compared with the thickness of the copper-brain spatula t_{Cu} for the maximum bending strain ε_m = 1- 4 % are shown in table 2. If the thickness shown in table 2 is used, the SMA-brain spatula can be bent by the same force as that required for the existing copper-brain spatula. As can be seen from table 2, the maximum bending strain ε_m affects slightly on the ratio of t_{SMA} to t_{Cu}. If the thicknesses of the rolled- and the cast-SMA brain spatula are 1.88 times and 1.16 times as large as that of the copper-brain spatula, respectively, the SMA-brain spatula can be bent by the same force as the existing copper-brain spatula.

4.3 Shape of SMA-brain spatula

Based on the discussion in the sections 4.1 and 4.2, it has been clarified that, if the length and the width of the rolled-SMA brain spatula are the same as those of the existing copper-brain spatula and the thickness of the rolled-SMA brain spatula is 1.3 times as large as that of the copper one, the rolled-SMA brain spatula can hold the same bending rigidity and can be bent easily by the smaller force than the copper one. If the thickness of the cast-SMA brain spatula is 1.2 times as large as that of the copper one, the cast-SMA brain spatula can hold the same bending rigidity and can be bent by almost the same force as the existing copper one.

Table 2. Thickness of SMA-brain spatula compared with that of copper-brain spatula t_{Cu} required to bend both spatulas by the same force for various maximum bending strains

Maximum bending strain	**Thickness of Rolled SMA**	**Thickness of Cast SMA**
$\varepsilon_m = 1$ %	**1.87** t_{Cu}	**1.17** t_{Cu}
$\varepsilon_m = 2$ %	**1.88** t_{Cu}	**1.16** t_{Cu}
$\varepsilon_m = 3$ %	**1.88** t_{Cu}	**1.16** t_{Cu}
$\varepsilon_m = 4$ %	**1.88** t_{Cu}	**1.16** t_{Cu}

5. Bending fatigue properties

5.1 Fatigue life in alternating-plane bending

The relationships between the maximum bending strain ε_m and the number of cycles to failure N_f obtained from the alternating-plane bending fatigue test for the rolled SMA, the cast SMA and the copper are shown in figure 9.

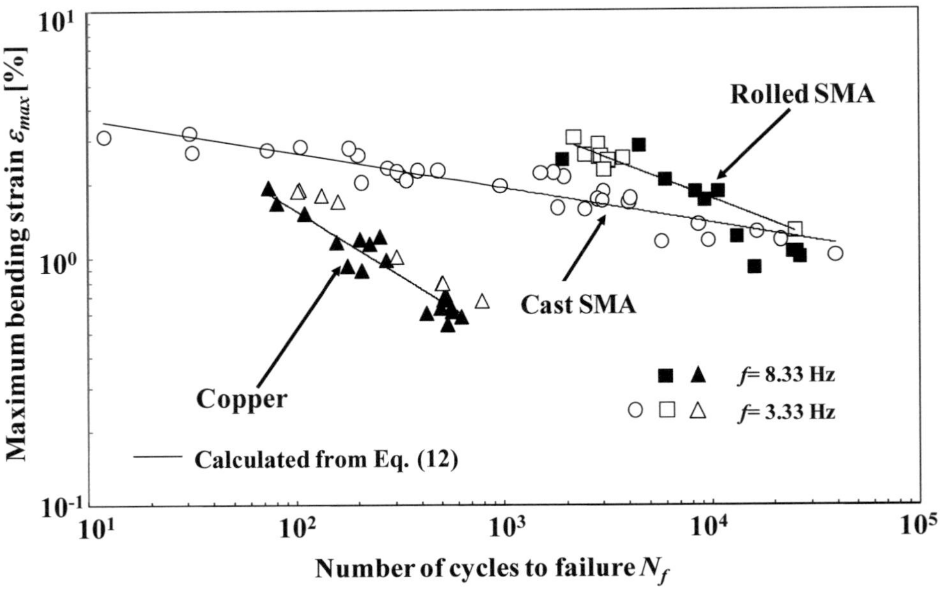

Fig. 9. Fatigue life curves of copper, rolled SMA and cast SMA used for brain spatula in alternating-plane bending

In figure 9, the larger the ε_m, the smaller the N_f for all materials. The fatigue life of the rolled SMA is longer than that of the copper by 100 times. The fatigue life of the cast SMA is longer that of the copper by 100 times in the region of small ε_m and by 10 times in the region of large ε_m. The fatigue life of the copper is several hundred cycles at ε_m = 1 %. This fatigue life is almost the same as that in the low-cycle fatigue of normal metals [10]. In the case of the copper, the yield strain is caused by the slip of crystals and the plastic deformation occurs repeatedly, resulting in the short fatigue life. In the case of the SMAs, the yield strain is caused not by the permanent slip but by the recoverable rearrangement of the M-phase, resulting in the long fatigue life. In the case of the cast SMA, as observed in figure 4, stress increases with an increase in strain, resulting in the shorter fatigue life with an increase in the maximum bending strain. The influence of the frequency on the fatigue life for all materials is not clear for f =3.33 Hz and 8.33 Hz. In the case of a TiNi SMA wire, the lower the frequency, the longer the fatigue life is [11]. The frequency of the SMA-brain spatula in the practical use should be lower than f =3.33 Hz and 8.33 Hz. Therefore, the fatigue life in the practical use will be longer than that obtained in the present study.

In figure 9, the fatigue life curves of all materials in the low-cycle fatigue region can be approximated by the straight lines. Since the relationships between ε_m and N_f can be expressed by the straight lines on the logarithmic graph, the relationship can be expressed by a power function similar to the fatigue life for the TiNi SMA wires and tubes [12] as follows

$$\left.\begin{array}{ll} \varepsilon_m \cdot N_f^{\ \beta} = \alpha & \\ & \\ \beta = 0.58,\ \alpha = 25\% & : \text{Copper} \\ \beta = 0.41,\ \alpha = 67\% & : \text{Rolled SMA} \\ \beta = 0.14,\ \alpha = 5\% & : \text{Cast SMA} \end{array}\right\} \qquad (12)$$

where α and β denote ε_m in $N_f = 1$ and the slope of the log ε_m - log N_f curves, respectively. The calculated results by Eq. (12) are shown by the solid lines in figure 9. As can be seen, the overall inclinations are well approximated by the solid lines.

5.2 Fatigue life in pulsating-plane bending

The relationships between the maximum bending strain ε_m and the number of cycles to failure N_f obtained from the pulsating-plane bending fatigue test for the cast SMA, the rolled SMA and the copper are shown in figure 10.

In figure 10, the larger the ε_m, the smaller the N_f for all materials. The fatigue life of the rolled SMA is longer than that of the copper by 100 times. The fatigue life of the cast SMA is longer that of the copper by 40 times. The fatigue life of the copper is two thousands cycles at ε_m = 2 %. In the case of the cast SMA, as observed in figure 4, the yield stress is higher than that of the rolled SMA, resulting in the shorter fatigue life than the rolled SMA.

In figure 10, the fatigue life curves of all materials in the low-cycle fatigue region can be approximated by the straight lines. The relationship can be expressed by a power function similar to Eq. (12) as follows

$$\left.\begin{array}{ll} \varepsilon_m \cdot N_f^{\ \beta} = \alpha & \\ & \\ \beta = 0.35,\ \alpha = 26\% & : \text{Copper} \\ \beta = 0.29,\ \alpha = 49\% & : \text{Cast SMA} \\ \beta = 0.26,\ \alpha = 43\% & : \text{Rolled SMA} \end{array}\right\} \qquad (13)$$

The calculated results by Eq. (13) are shown by the solid lines in figure 10. As can be seen, the overall inclinations are well approximated by the solid lines.

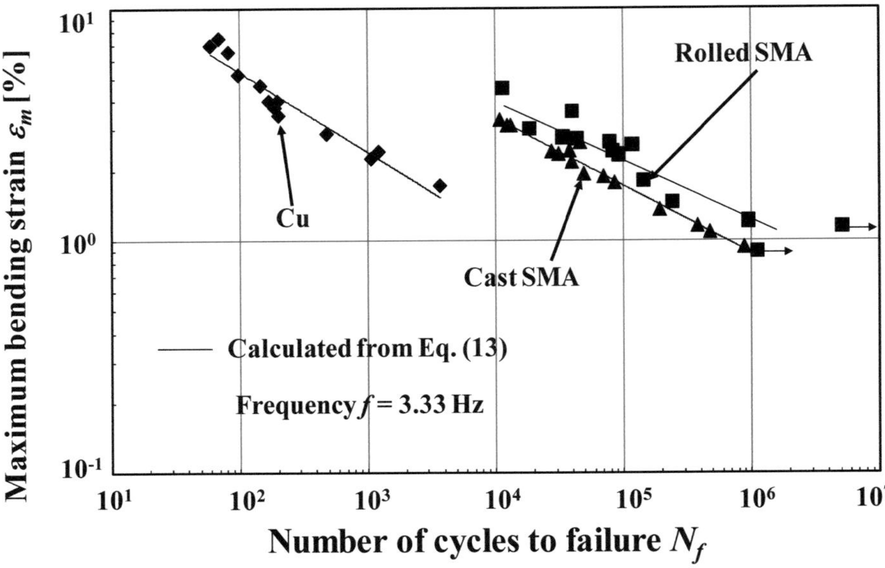

Fig. 10. Fatigue-life curves of copper, rolled SMA and cast SMA used for brain spatula in pulsating-plane bending

5.3 Comparison of fatigue life

The relationships between the maximum bending strain ε_m and the number of cycles to failure N_f obtained from the alternating- and pulsating-plane bending fatigue tests for the rolled and cast SMAs are shown in figures 11 and 12, respectively. In figures 11 and 12, the fatigue life in alternating-plane bending is shorter than that in the pulsating-plane bending. The dissipated work in each cycle of the alternating-plane bending is larger than that in the pulsating-plane bending. The fatigue damage is greater in the alternating-plane bending, and therefore the fatigue life in the alternating-plane bending is shorter than that in the pulsating-plane bending. The influence of the dissipated work W_d on the fatigue life between the alternating- and pulsating-plane bendings can be explained as follows. The stress-strain diagram and the dissipated work W_d at each cycle in the pulsating- and alternating-plane bendings are schematically shown in figure13. In figure 13, it is assumed that the yield stress σ_M is constant in the stress-plateau region and takes the same value in tension and compression. The area surrounded by the hysteresis loop denotes the dissipated work W_d per unit volume in each cycle. The dissipated work W_d is expressed by the following equation in the pulsating-plane bending

$$W_d = 2\sigma_M(\varepsilon_m - \frac{2\sigma_M}{E}) \tag{14}$$

and that in alternating-plane bending

$$W_d = 2\sigma_M(2\varepsilon_m - \frac{2\sigma_M}{E}) \tag{15}$$

where E represents the elastic modulus. It can be seen from Eqs. (14) and (15) that W_d increases in proportion to both σ_M and ε_m. The values of the dissipated W_d for ε_m=3% by using Eqs. (14) and (15) are shown in Table 3. As can be seen in Table 3, the value of W_d for the cast SMA in the case of the alternating-plane bending is very large compared to that of the pulsating-plane bending. Therefore, in the case of the cast SMA, the fatigue life of the alternating-plane bending is short for large maximum bending strain ε_m.

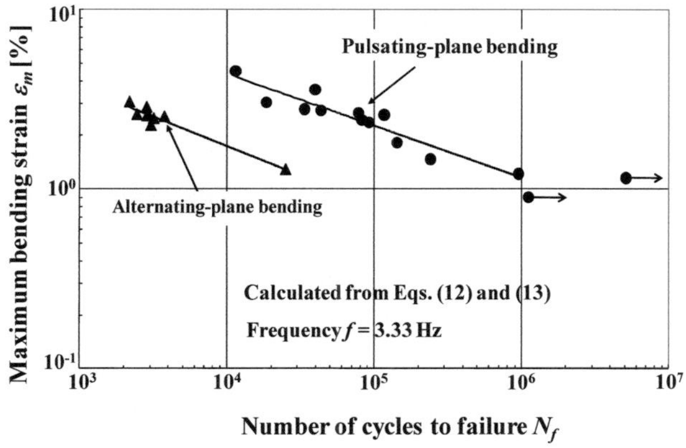

Fig. 11. Relationship between maximum bending strain and number of cycles to failure of rolled SMA in pulsating-plane and alternating-plane bending

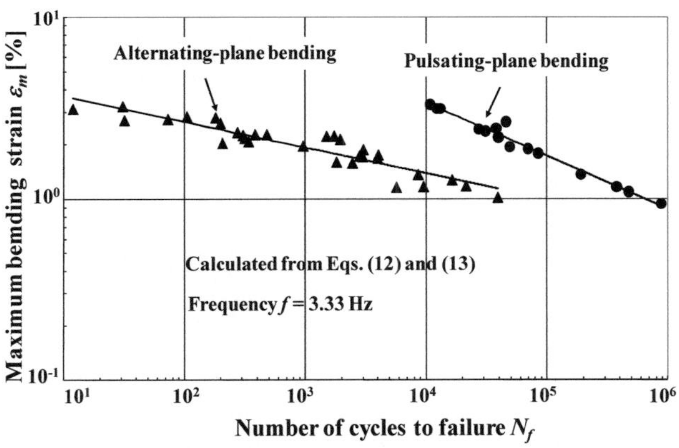

Fig. 12. Fatigue-life curves of cast SMA used for brain spatula in pulsating-plane and alternating-plane bending

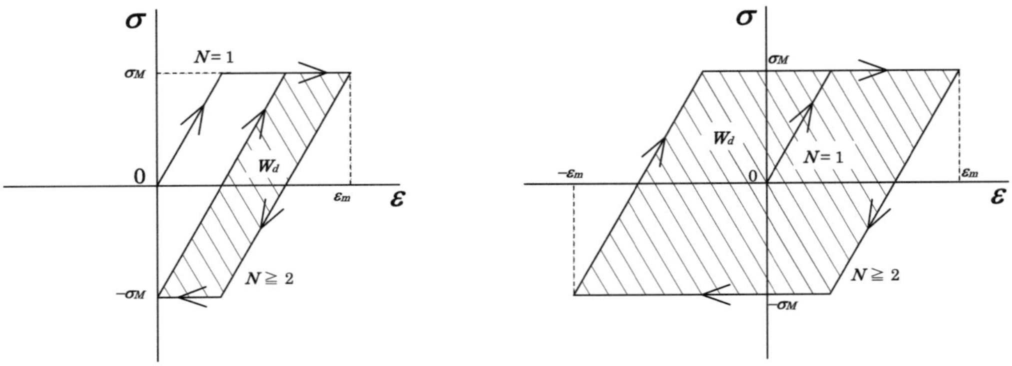

(a) Pulsating-plane bending

(b) Alternating-plane bending

Fig. 13. Stress-strain diagram and dissipated work W_d at each cycle in pulsating-plane bending and alternating-plane bending

Table 3. Values of dissipated work W_d in pulsating- and alternating-plane bendings for $e_m = 3\%$

Materials	Dissipated work W_d (MJ/m^3)	
	Pulsating bending	Alternating bending
Rolled SMA	3.6	7.7
Cast SMA	8.0	18
Copper	12	26

5.4 Temperature rise under cyclic bending

The surface element of the specimen is subjected to the dissipated work in each cycle as discussed in section 5.3. Therefore, temperature of the specimen increases during cyclic bending. The temperature on the surface of the central part of the rolled SMA was measured by using a thermocouple in the case of the alternating-plane bending at a frequency of f =3.33 Hz. The relationship between the temperature rise ΔT and time is shown in figure 14. The temperature increases rapidly in the early 20 s. After 20 s, the amount of generated heat and released heat matches each other and ΔT saturates at a certain value.

The relationship between the saturated temperature rise ΔT_s and the maximum bending strain ε_m for the rolled and cast SMAs at f = 3.33Hz is shown in figure 15. The saturated temperature rise ΔT_s increases in proportion to the maximum bending strain ε_m. The value of ΔT_s for the cast SMA is a little larger than that for the rolled SMA, and is about 30K at ε_m =3%. The yield stress σ_M increases in proportion to temperature rise ΔT by 5MPa/K for TiNi SMA. Therefore, if ΔT = 30K, the yield stress increases by 150MPa. This means the fact that the fatigue damage is large in the case of alternating-plane bending for the large ε_m. In the case of practical use of the brain spatula, the frequency is low and the temperature rise is small, resulting in the longer fatigue life than that obtained in the present study.

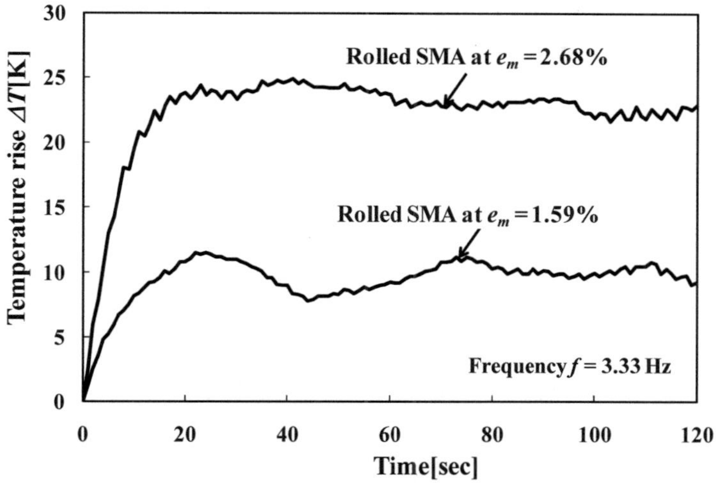

Fig. 14. Relationships between temperature rise and time for rolled SMA in alternating-plane bending at f =3.33 Hz

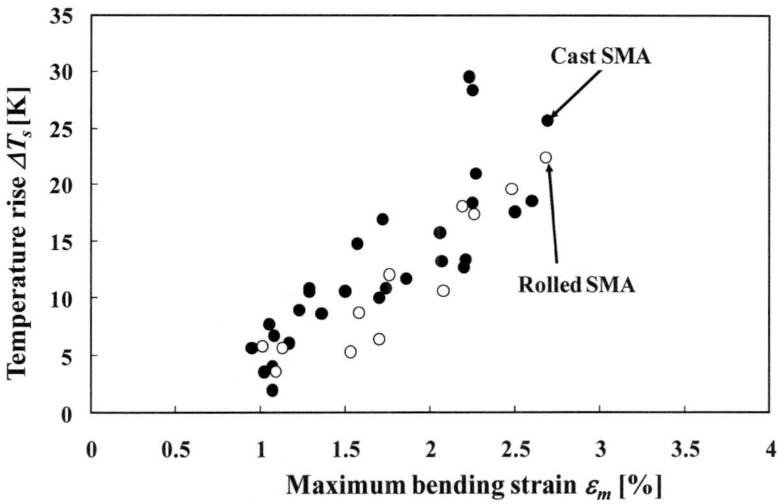

Fig. 15. Relationships between saturated temperature rise and maximum bending strain in alternating-plane bending at f =3.33 Hz

5.5 Observation of Fracture Surface.

The SEM photograph of the fracture surface for the rolled SMA in the case of the pulsating-plane bending at ε_m = 2.66% and N_f = 78081 cycles is shown in figure 16. As can be seen, the fatigue crack nucleates at one corner and grows into the central part. The crack growth rate is higher along the surface since the stress of the surface element is higher than that of the central element. This fatigue crack growth is the same inclination as that observed in the case of the alternating-plane bending for the rolled SMA.

The SEM photograph of the fracture surface for the cast SMA in the case of the pulsating-plane bending fatigue at ε_m = 2.41% and N_f = 27417 cycles is shown in figure 17. The fatigue crack nucleates at one corner and grows into the central part. The overall inclination of the fatigue crack growth is similar to that for the rolled SMA shown in figure 16.
The SEM photograph of the fracture surface for the cast SMA in the case of the alternating-plane bending fatigue at ε_m = 1.74% and N_f = 4052 cycles is shown in figure 18. The casting defect is observed in the neighbourhood of the surface and the fatigue crack grows from this casting defect. If the casting defect exists on the surface element, the fatigue crack starts in the early cycle. Therefore the development of casting method without defects is the future subject.

Fig. 16. SEM photograph of fracture surface for rolled SMA in pulsating-plane bending fatigue

Fig. 17. SEM photograph of fracture surface for cast SMA in pulsating-plane bending fatigue

Fig. 18. SEM photograph of fracture surface for cast SMA in alternating-plane bending fatigue

6. Conclusions

In order to develop the SMA-brain spatula, the mechanical characteristics of the TiNi cast- and rolled-SMAs and the copper used for the brain spatula were compared based on the tensile deformation properties, and the characteristics of the SMA-brain spatula were discussed. The fatigue properties of these materials were also investigated by the pulsating-

and alternating-plane bending fatigue tests. The results obtained can be summarized as follows.

Based on the yield stress and the modulus of elasticity of the copper and the TiNi SMAs, the bending deformation properties of the SMA-brain spatula were estimated by assuming the condition to use the brain spatula as the bending of the strip cantilever. With respect to the SMA-brain spatula for the same length and width as the existing copper one, if the thickness of the conventional rolled-SMA spatula is 1.3 times as large as that of the existing copper-brain spatula, the SMA spatula can hold the same bending rigidity and can be bent by smaller force than the existing copper one. If the thickness of the new cast-SMA spatula is 1.2 times as large as that of the existing-copper spatula, the SMA spatula can hold the same bending rigidity and can be bent by almost the same force as the existing copper one.

With respect to the alternating- and pulsating-plane bending fatigue, the fatigue life of both the copper and the SMAs in the region of low-cycle fatigue is expressed by a power function of the maximum bending strain. The fatigue life of the conventional rolled SMA and the new cast SMA is longer than that of the existing copper. The fatigue life of the new cast and rolled SMAs in the pulsating-plane bending is longer than that in the alternating-plane bending.

The above mentioned characteristics of the SMA-brain spatula obtained in this study will be substantially applied to the development not only for the brain spatula but also for other retractors and instruments used in other surgery operations.

Acknowledgment

The experimental work of this study was carried out with the assistance of the students in Aichi Institute of Technology, to whom the authors wish to express their gratitude. The samples of the existing copper-brain spatula were supplied by Mizuho Co., Ltd., to whom the authors wish to express their gratefulness. The authors also wish to extend their thanks to the Basic Research (C) of Grant-in-Aid of Scientific Research supported by Japan Society for Promotion of Sciences and the Program of Support for Advanced Corporate Networking supported by Chubu Bureau of Economy, Trade and Industry, METI for their financial support.

7. References

[1] Funakubo H, ed. 1987, *Shape memory alloys*, Gordon and Breach Science Pub., 1-60

[2] Duerig T W, Melton K N, Stockel, D and Wayman C M eds., 1990, *Engineering Aspects of Shape Memory Alloys*, Butterworth-Heinemann, 1-35

[3] Saburi, T, ed. 2000, *Shape Memory Materials,* Trans Tech Pub., Switzerland.,

[4] Chu, Y. Y. and Zhao, L. C., eds., 2002, *Shape Memory Materials and Its Applications,* Trans Tech Pub., Switzerland.

[5] Otsuka, K. and Wayman, C. M., eds., 1998, *Shape Memory Materials,* Cambridge University Press, Cambridge.

[6] Yoshimi Y, Kitamura K, Tokuda M, Inaba T, Asai J and Watanabe Y, 2008, *Ti-Ni Shape Memory Alloys Precision Casting Products and Its Process,* Proc. Int. Conf. Shape Memory and Superelastic Tech., 387-396.

[7] Tobushi H, Tanaka K, Kimura K, Hori T and Sawada T, 1992, *JSME Inter. J.,* 35-3 278-284.

[8] Furuichi Y, Tobushi H, Ikawa T and Matsui R, 2003, *J. Materials: Design and Applications* 217-Part L 93-99.

[9] Tobushi, H., Okumura, K., Nakagawa, K. and Takata, K., Fatigue Properties of TiNi Shape Memory Alloy, Trans. Mater. Res. Soc. Jap., 26-1, 2001, pp. 347-350.

[10] Shigley J. E. and. Mischke C. R., 1989, *Mechanical Engineering Design, 5th ed.*, McGraw-Hill, New York, 270-274.

[11] Tobushi H, Nakahara T, Shimeno Y and Hashimoto T, 2000, *Low-Cycle Fatigue of TiNi Shape Memory Alloy and Formulation of Fatigue Life*, Trans. ASME, J. Eng. Mater. Tech., 122 186-191

[12] Matsui, R., Tobushi, H., Furuichi, Y. and Horikawa, H., Tensile Deformation and Rotating-Bending Fatigue Properties of a Highelastic Thin Wire, a Superelastic Thin Wire, and a Superelastic Thin Tube of NiTi Alloys, Trans. ASME, J. Eng. Mater. Tech., 126, (2004), pp. 384-391.

4

Hysteresis behaviour and modeling of SMA actuators

Hongyan Luo[1,2], Yanjian Liao[1], Eric Abel[2],
Zhigang Wang[2] and Xia Liu[1]
[1]*Chongqing University*
P.R. China
[2]*University of Dundee*
UK

1. Introduction

In recent years, the use of shape memory alloy (SMA) as a key component in diverse actuation applications has attracted more interests, especially in the field of mechatronics and medical instruments (Wolfe et al, 2005; Wong et al, 2007; Gupta et al, 2009, Okamura et al, 2009, De Sars et al, 2010). The positive features of good reliability, high energy density, design simplicity, compactness in configuration and quiet operation, point to SMA being a promising candidate for actuator. However, great difficulties are always encountered in the precise control of the systems incorporating them, due mainly to the nonlinearities of the complex hysteresis associated with the shape memory effect. These nonlinearities must be considered and dealt with properly, since they may excite unwanted dynamics that lead in the best scenario to a deteriorated system performance and in the worst scenario to an unstable dynamic system.

One effective method to compensate for such hysteresis nonlinearities is to involve a model in the control scheme that is able to describe the complex nonlinear behaviour of SMA actuators and accordingly give reliable predictions of the system response. In this case, the crucial part of the development lies in accurate modelling of the actual hysteresis nonlinearity. From the viewpoint of control, such a hysteresis model should characterize the nonlinearities with sufficient accuracy, be amenable to a compensator design for actuator linearization and be well-suited for real-time applications. Therefore, the usual constitutive models (Bhattacharyya & Lagoudas, 1997; Matsuzaki & Naito, 2004; Popov & Lagoudas, 2007; Wang & Dai, 2010), derived from thermodynamical or micromechanical principles, are immediately discarded for this purpose, owing to the mathematical complexity involved and non-availability of the microstructural material parameters.

This chapter starts with the description of the hysteresis behaviour of SMA actuators. Following the analysis of its hysteresis characteristics, a phenomenological model based on the theory of hysteresis operator, referred to as MKP model, is proposed and its inverse model is deduced with the aim to provide a more appropriate choice for the modelling of the hysteretic system incorporating SMA actuator for control applications. The computer simulations are also presented to verify this model.

2. Hysteresis characteristics of SMA actuators

The hysteresis behaviour of SMA actuators associated with shape memory effect is characterized by the fact that there are multiple possible outputs (i.e. strain or displacement) with respect to a given input (i.e. temperature). At any time, the specific output is determined by the current input and the previous inputs (i.e. input history). This type of relationship between input and output, given the name of hysteresis, is a byproduct of phase transitions that take place in SMA, and is caused by the internal friction generated by the movement of the austenite-martensite interface and by the creation of structural defects within the alloy structure (Wayman & Duerig, 1990). This strain-temperature hysteresis is present in all systems incorporating SMA actuators and is an important consideration in their control.

To give a demonstration of such hysteresis behaviour, some sample curves were measured on a NiTi tension spring under a load by alternately heating and cooling it between the temperature lower than its martensite finish temperature (Mf) and the temperature higher than its austenite finish temperature (Af). Fig.1 gives a typical temperature-displacement curve where the points at which the temperature gradient changes its sign, known as return points, are marked by number. The curve shows a major loop enclosing a group of minor loops.

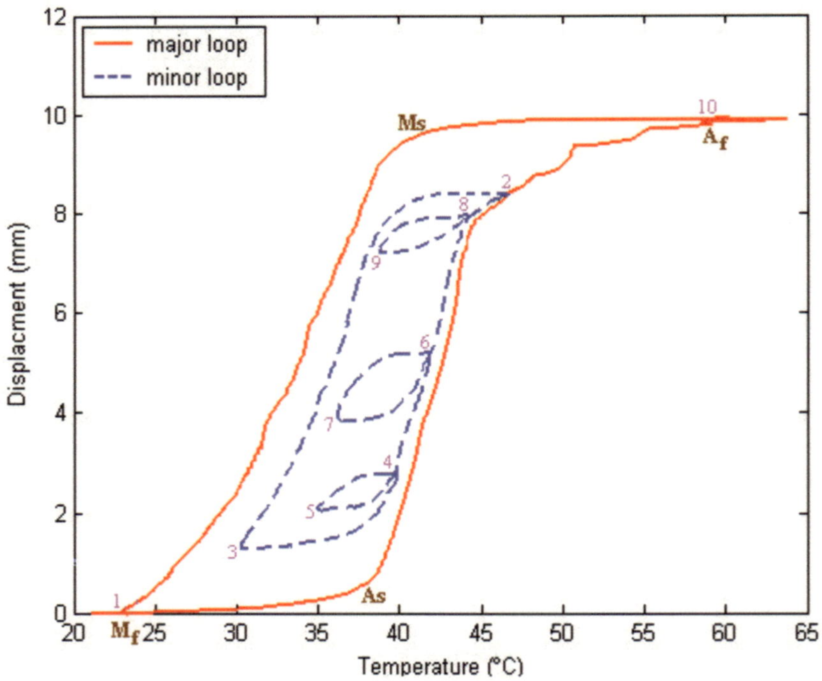

Fig. 1. A typical hysteresis curve of an SMA actuator by the example of a NiTi tension spring

The major loop is the full-scale strain trajectory corresponding to a complete actuation cycle of an SMA actuator, which is achieved by heating and cooling the actuator throughout its overall transformation temperature range defined by Mf (lower limit) and Af (upper limit). The minor loop corresponds to the partial strain excursion within the strain range of the SMA actuator caused by heating and cooling the actuator through part of its transformation temperature range, i.e. the incomplete actuation. Although the specific shape and width of hysteresis loop are related to alloy composition and the fabrication process (Van Humbeeck & Stalmans, 1998), there seems to be general agreement in the literature on the main features of SMA hysteresis response observed experimentally, such as closed loops between two return points, path dependency, self-containment and return point memory (Ŝittner et al, 2000).

Noticeably, for the same SMA actuator, the envelope of its hysteresis curve, i.e. the major loop, does not keep constant, but varies with the applied loading and the amount of strain prior to actuation, as illustrated in Fig. 2. It can be seen that both the increasing mass applied to the SMA actuator and the increasing compressed length (i.e. pre-deformation), cause the major loop of the SMA actuator to expand vertically. Inside the major loop, any minor loop, which is formed by reversing the actuation direction of SMA actuator in the state of mixed martensite and austenite, is mainly determined by the internal microstructure or rather the interaction of co-existing martensite and austenite, depending on the temperature history.

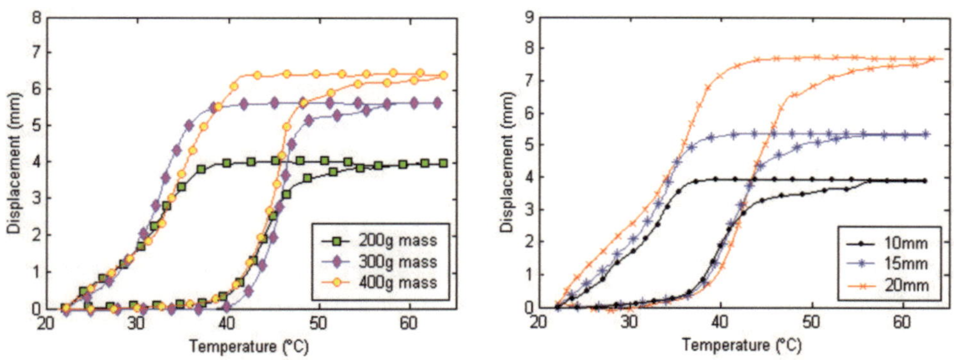

(a) Major loops for different masses (b) Major loops for different pre-deformation

Fig. 2. Influence of applied loading and pre-deformation on hysteresis behaviour of SMA actuators

On one hand, the common characteristics in the hysteresis behaviour allow the possibility of a purely phenomenological model to generally represent the hysteresis response of various SMA actuator based systems, regardless of the type of SMA actuator involved and the underlying physics. On the other hand, the complex hysteresis behaviour, which is influenced by both external and internal factors, makes it a challenge to implement algorithms for the hysteresis modelling in dynamic service environments, since a great deal of parametric uncertainties are involved. In particular, the modelling issue becomes much more complicated in the application of SMA actuators under controlled actuation, where they typically work in partial cycles and thus many minor loops would be encountered rather than a single major loop in the case of on-off applications. Furthermore, for real-time control applications, the complexity of algorithm implementation and computational burden associated with the phenomenological model are especially concerned.

3. MKP hysteresis model

The MKP hysteresis model proposed here is actually a mutation of the classical Preisach model, which is a well-known phenomenological hysteresis model and based on the general theory of hysteresis operators (Krasnoselskii & Pokrovskii, 1989). The rest of this section only covers the main principles of this theory. A more complete description is available elsewhere (Mayergoye, 1991; Krasnoselskii et al, 1994; Webb & Lagoudas, 1998).

3.1 Principle and mathematical formalism

The concept of hysteresis operator modelling was first proposed in 1935 by the physician Preisach as a restricted physical representative for ferromagnetic hysteresis phenomenon, called Preisach model. About thirty years later, its nature in phenomenology was noticed and the mathematical properties were investigated by the mathematicians (Krasnoselskii & Pokrovskii, 1989), who separated this model from its physical meaning and represented it in a purely mathematical form as follows:

$$Y = H[T(t)] = \iint_{(\alpha,\beta)\in S} \mu(\alpha,\beta)\gamma_{\alpha\beta}(T,\xi_{\alpha\beta})d\beta d\alpha \tag{1}$$

Here $T(t)$ is the input (i.e. temperature for SMA actuator), $\hat{\gamma}_{\alpha\beta}$ is an elementary hysteresis operator function that gives the current states of the hysteresis operators in response to the input variation, the variable $\xi_{\alpha\beta}$ represents the previous state of the hysteresis operator, $\mu(\alpha,\beta)$ is the density distribution function, also called Preisach function, to define the weighting values for these operators, which captures characteristics of the specific hysteresis system, and S is the Preisach plane, i.e. the region over which hysteresis occurs, defined by the input region as follows:

$$S = \{(\alpha,\beta) : T_{\min} \le \alpha \le \beta, T_{\min} \le \beta \le T_{\max}\}$$

The integration takes the past input history into account to determine the current output.
For numerical implementation, a finite-dimensional approximation of equation (1) is deduced as

$$Y = H(T) = \sum_{j=1}^{K}\sum_{i=1}^{j} \gamma_{s_{ij}}(T,\xi_{s_{ij}})\mu_{s_{ij}} \tag{2}$$

by uniformly dividing the Preisach plane into a mesh grid (i.e. the discrete Preisach plane) of $K \times K$ with an interval of $\Delta T = (T_{\max} - T_{\min})/(K-1)$, as shown in Fig.3. Then the number of grid points representing the Preisach plane is given by $N = K \times (K+1)/2$. In the discrete Preisach plane, there exists a hysteresis operator $\gamma_{s_{ij}}(T,\xi_{s_{ij}})$ and a weighting value $\mu_{s_{ij}}$ for each grid point $s_{ij} = (\alpha_i, \beta_j)$, where

$$\alpha_i = T_{\min} + (i-1)\Delta T\ ,\ \beta_j = T_{\min} + (j-1)\Delta T \tag{3}$$

for $j = 1,2,\cdots,K$, $i = 1,2,\cdots,j$. As $K \to \infty$, the discretization becomes finer and finer, approaching the continuous Preisach plane. In the discrete Preisach plane, there exists a hysteresis operator $\gamma_{s_{ij}}(T,\xi_{s_{ij}})$ and a weighting value $\mu_{s_{ij}}$ for each grid point s_{ij} . In other words, each operator is identified by only one particular grid point, which is uniquely defined by a pair of α and β coordinates.

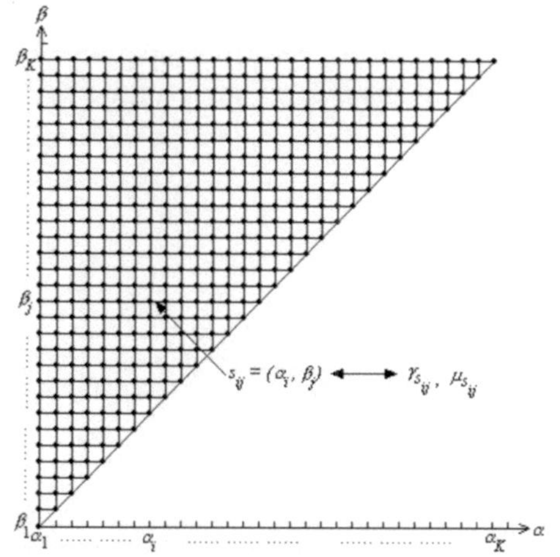

Fig. 3. Discretization of Preisach plane

The numerical approximation (see equation 2) apparently indicates that hysteresis operator modelling can be represented by a straightforward structure, analogous to the parallel connections of a series of qualitatively similar elements, as illustrated by the block diagram in Fig.4.

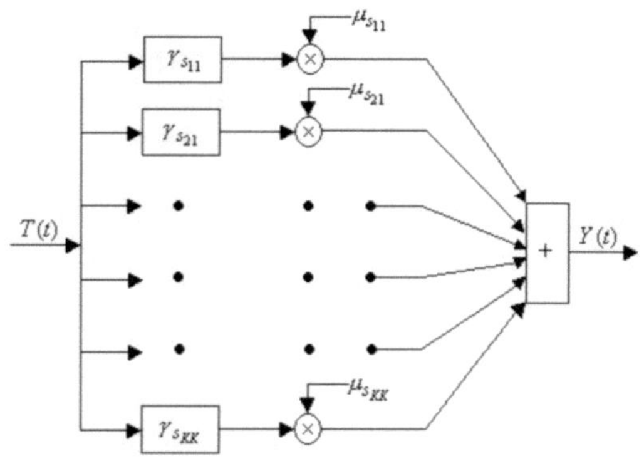

Fig. 4. Block diagram of hysteresis operator theory

Clearly, the above formalism presents a general mathematical tool for hysteresis modelling, which can be adjusted to capture the input-output features of the different hysteresis systems by the appropriate selection of the elementary hysteresis operator and of the density distribution function. This has allowed its applications to been extended from a ferromagnetic material to smart materials such as piezoceramic and SMA (Dickinson et al, 1996; Hughes & Wen, 1997).

3.2 MKP hysteresis operator

The building block of this hysteresis modelling approach is an elementary hysteresis operator, which is non-complicated hysteresis nonlinearity with a simple mathematical structure characterized by one or more parameters. The different types of elementary hysteresis operators in use account for the differences in the models based on the theory of hysteresis operator. The typical elementary hysteresis operators are relay operator and KP operator, which are used in the classical Preisach model and KP model respectively. The examples of these two types of hysteresis operator and MKP operator for use in MKP model, corresponding to a grid point $s_{ij} = (\alpha_i, \beta_j)$ in the discrete Preisach plane, are given in Fig. 5.

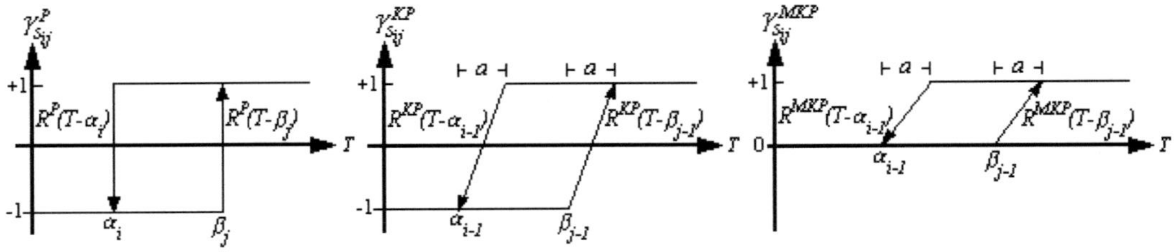

(a) Relay operator (b) KP operator (c) MKP operator

Fig. 5. Three types of elementary hysteresis operator

The relay operator is the simplest operator with only two output states, and characterized by a step transition between -1 and +1. The KP operator is a continuous version of relay operator, developed by Krasnoselskii and Pokrovskii (Krasnoselskii & Pokrovskii 1994), which can exist at any value in the closed interval of $[-1,+1]$ rather than at the two extreme ends. Due to this continuity, the KP model can describe a real hysteresis I/O relationship more closely than can the Preisach model. It is useful to point out that the selection of -1 and +1 as the extreme state values for both the relay operator and the KP operator arises from the initially physical motivation of the hysteresis operator theory that each elementary hysteresis operator represents one magnetic dipole changing between the negative magnetic polarity and the positive magnetic polarity (Hughes & Wen, 1997). This boundary condition is not compulsory for the definition of elementary hysteresis operator in terms of the mathematical meaning of this modelling theory, which need not be conformed to especially when it is used for modelling other materials. Therefore, the MKP operator is given here by simply halving the output range of the KP operator to $[0,+1]$.

Referring to Fig.5(c), the ridge function of a MKP operator is defined by

$$\begin{cases} 0 & x < 0 \\ r(x) = \dfrac{x}{a} & 0 \le x \le a \\ 1 & x > a \end{cases} \tag{4}$$

where $x = T - \alpha_{i-1}$ for the left bounding curve, $x = T - \beta_{j-1}$ for the right bounding curve. The rise interval of input a is set as ΔT, over which the operator changes value between 0 and 1. The mathematical representation, which describes the action of a MKP operator is given by

$$\gamma_{s_{ij}}(T,\xi_{s_{ij}}) = \begin{cases} \max\left\langle \xi_{s_{ij}}, r(T-\beta_{j-1}) \right\rangle & \dot{T} > 0 \\ \min\left\langle \xi_{s_{ij}}, r(T-\alpha_{i-1}) \right\rangle & \dot{T} < 0 \\ \xi_{s_{ij}} & \dot{T} = 0 \end{cases} \tag{5}$$

where $\dot{T}$ represents the direction of temperature change and the memory term (i.e. $\xi_{s_{ij}}$) for a specific MKP operator is always updated by the instantaneous value of $\gamma_{s_{ij}}$.

3.3 Advantages of MKP model

Fig.6(a-d) shows how the Preisach plane evolves and how MKP operators respond to input changes. When input is lower than $T_{\min}$, all the MKP operators are 'off' at the minimum output state. As input increases, a horizontal line slides along the positive β direction with the operators just above it as a horizontal boundary, below which all the operators are switched to the maximum output state. Following the decreasing input, a vertical line sweeps in the negative α direction with the operators right next to this line as a vertical boundary, right to which all the operators are changed back to minimum output state. For a sequence of input alternating between $T_{\min}$ and $T_{\max}$, a staircase interface line (i.e. boundary) is created. From the viewpoint of a geometrical interpretation, the Praisach plane at each time instant is divided into two parts: $S_{\max}$ where the operators are at the maximum output state (dark grey), $S_{\min}$ where the operators are at the minimum output state (light grey), by a boundary (S_{bon}) where the operators are at the intermediate state. By comparing the evolution of Preisach plane regarding MKP model, Preisach model and KP model, the advantages of MKP model over the other two can be revealed. Firstly, owing to the minimum output state of 0, the MKP operators in the area $S_{\min}$ make no contribution to the output of the MKP model and need not be taken into account. As a result, the MKP model is more efficient in calculation than KP model and Preisach model, which makes it more suitable for real-time control applications. This advantage is more outstanding, when the discretization of Preisach plane becomes finer and the number of parameters is vast. Secondly, for a SMA based system, its output is usually measured in terms of displacement relative to its initial position, so that its output range is defined as $[0, Y_{\max}]$. Evidently, this is well matched by the output range of MKP model, i.e. $[0, H(T_{\max})]$. These features allow MKP model to be a more appropriate choice for the modelling of the hysteretic system incorporating SMA actuator.

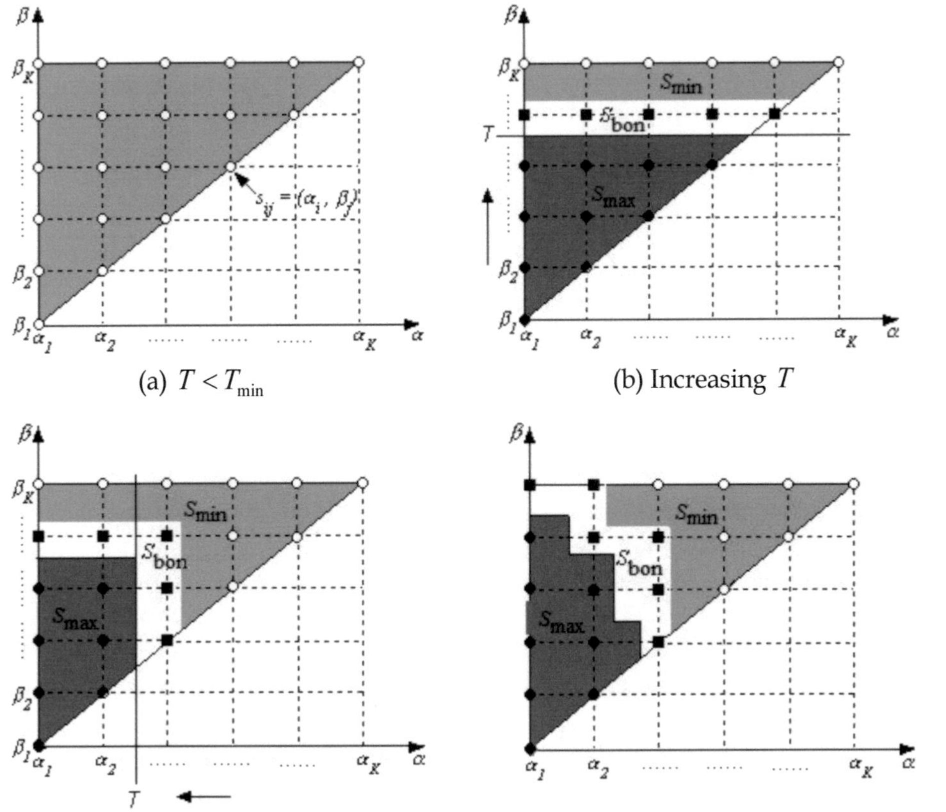

(a) $T < T_{\min}$ (b) Increasing T

(c) Decreasing T (d) Alternating T between $T_{\min}$ and $T_{\max}$

Fig. 6. Evolution of Preisach plane

(o: 0 valued operator; ■: (0,1) valued operator; •: 1 valued operator)

4. Inverse MKP hysteresis model

In fact, the MKP model that gives an input-output mapping cannot directly serve the control applications, since the hysteresis system is often required to follow a reference trajectory, a situation that the desired output is always known, while the corresponding input is unknown. Therefore, it is exactly the inverse MKP model with the capability to predict an input corresponding to a desired output can act as a compensator in control schemes to tackle the hysteresis nonlinearity.

The inverse MKP hysteresis model, represented by $H^{-1}(Y)$, can be easily formulated, after understanding how MKP operators respond to input changes. Let the current input be $T_c \in (\beta_{p-1}, \beta_p]$ and the corresponding output be Y_c. Consider the case of increasing output first.

If the next target output is $Y_d > Y_c$, the goal is to determine T_d so that $Y_d = H(T_d)$. According to the monotonicity, $Y_d > Y_c$ indicates $T_d > T_c$. Therefore, the MKP operators should be evolved along the positive β axis. By starting the direct calculation of the forward MKP hysteresis model from $T=\beta_p$ in increment of ΔT until $H(\beta_m)\geq Y_d$, it can be determined that $T_d \in (\beta_{m-1},\beta_m]$. Recalling the previously introduced geometrical interpretation of MKP model (see section 3.3), the final link of the interface to be created by T_d in the Preisach plane is a horizontal one and appears on the row m. Then only the output states of the MKP operators locating on this row may be related to T_d, while all the MKP operators locating on the rows lower than m will have the maximum output states of +1 and those locating on the rows higher than m will keep the previous output states. Referring to the definition of MKP operator function (equation (4),(5)), this can be expressed analytically as

$$\gamma_{s_{ij}}(T_d,\xi_{s_{ij}})=\begin{cases} 1 & \text{for } j=1,2,\cdots,m-1 \text{ and } i=1,2,\cdots,j \\ \xi_{s_{ij}} & \text{for } j=m+1,m+2,\cdots,K \text{ and } i=1,2,\cdots,j \\ \max\left\langle \xi_{s_{ij}},\dfrac{T_d-\beta_{m-1}}{\Delta T}\right\rangle & \text{for } j=m \text{ and } i=1,2,\cdots,m \end{cases} \tag{6}$$

Then the calculation of Y_d can be expressed as

$$Y_d=\sum_{j=1}^{m-1}\sum_{i=1}^{j}\mu_{s_{ij}}+\sum_{j=m+1}^{K}\sum_{i=1}^{j}\xi_{s_{ij}}\mu_{s_{ij}}+\sum_{i=1}^{m}\gamma_{s_{im}}(T_d,\xi_{im})\mu_{s_{im}} \tag{7}$$

Clearly, the output states of the operators on the row m will change differently, depending on their previous output states and the exact value of T_d. For the most general discussion, assume the past input history causes a staircase line between β_{m-1} and β_m, and the operators on the row m have the previous output states as shown in Fig. 7.

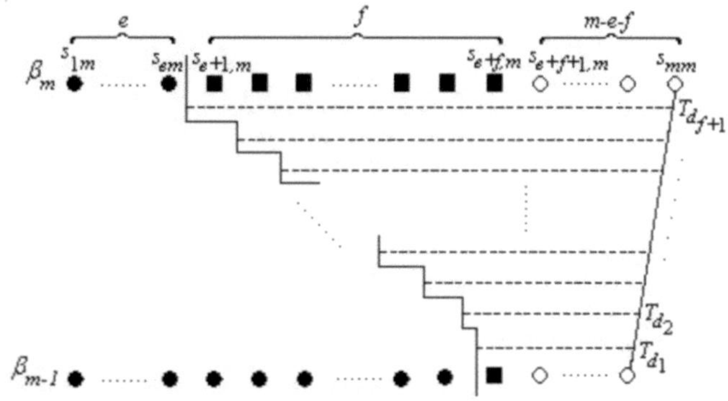

Fig. 7. Example of inverse MKP model for the increasing desired output

(o: 0 valued operator; ■: (0,1) valued operator; •: 1 valued operator)

Then the following results can be obtained:
For $i=1,2,\cdots,e$, since $\xi_{im}=1$, there exists

$$\gamma_{s_{im}}(T_d,\xi_{s_{im}})=\max\left\langle 1,\frac{T_d-\beta_{m-1}}{\Delta T}\right\rangle=1$$

For $i = e+f+1, e+f+2, \cdots, m$, since $\xi_{im} = 0$, there exists

$$\gamma_{s_{im}}(T_d, \xi_{s_{im}}) = \max\left\langle 0, \frac{T_d - \beta_{m-1}}{\Delta T} \right\rangle = \frac{T_d - \beta_{m-1}}{\Delta T}$$

For $i = e+1, e+2, \cdots, e+f$, since $0 < \xi_{im} < 1$, there are $f+1$ possible situations, namely,

i) $\gamma_{s_{im}}(T_d, \xi_{s_{im}}) = \xi_{s_{im}}$, for $i = e+1, e+2, \cdots, e+f$

ii) $$\gamma_{s_{im}}(T_d, \xi_{s_{im}}) = \begin{cases} \xi_{s_{im}} & \text{for } i = e+1, e+2, \cdots, e+f-1 \\ \dfrac{T_d - \beta_{m-1}}{\Delta T} & \text{for } i = e+f \end{cases}$$

iii) $$\gamma_{s_{im}}(T_d, \xi_{s_{im}}) = \begin{cases} \xi_{s_{im}} & \text{for } i = e+1, e+2, \cdots, e+f-2 \\ \dfrac{T_d - \beta_{m-1}}{\Delta T} & \text{for } i = e+f-1, e+f \end{cases}$$

…

$f+1$) $\gamma_{s_{im}}(T_d, \xi_{s_{im}}) = \dfrac{T_d - \beta_{m-1}}{\Delta T}$, for $i = e+1, e+2, \cdots, e+f$

This indicates that there are $f+1$ possible solutions for T_d. Let n be the number of the operators that are switched from the previous intermediate output states to the new ones in each case, i.e. $n = 0, 1, \cdots, f$. The equation (7) can be rewritten as

$$Y_d = \sum_{j=1}^{m-1}\sum_{i=1}^{j} \mu_{s_{ij}} + \sum_{j=m+1}^{K}\sum_{i=1}^{j} \xi_{s_{ij}}\mu_{s_{ij}} + \sum_{i=1}^{e} \mu_{s_{im}} + \sum_{i=e+1}^{e+f-n} \xi_{s_{im}}\mu_{s_{im}} + \sum_{i=e+f-n+1}^{m} \frac{T_d - \beta_{m-1}}{\Delta T}\mu_{s_{im}} \tag{8}$$

from which, the expression for T_d can be obtained as

$$T_d = \beta_{m-1} + \frac{\Delta T}{\sum\limits_{i=e+f-n+1}^{m} \mu_{s_{im}}} \times \left(Y_d - \sum_{j=1}^{m-1}\sum_{i=1}^{j} \mu_{s_{ij}} - \sum_{j=m+1}^{K}\sum_{i=1}^{j} \xi_{s_{ij}}\mu_{s_{ij}} - \sum_{i=1}^{e} \mu_{s_{im}} - \sum_{i=e+1}^{e+f-n} \xi_{s_{im}}\mu_{s_{im}}\right) \tag{9}$$

By replacing n with the value of $0, \cdots, f$ in turn in the above equation, all the possible solutions (i.e. $T_{d_1}, T_{d_2}, \cdots, T_{d_{f+1}}$) for T_d can be calculated and the exact value of T_d is determined as the minimum, namely

$$T_d = H^{-1}(Y_d) = \min\left\langle T_{d_1}, T_{d_2}, \cdots, T_{d_{f+1}} \right\rangle \tag{10}$$

When the desired output is decreasing, the similar deduction procedure is carried out. For the target output $Y_d < Y_c$, the required input should be $T_d < T_c$. Then the MKP operators will be evolved along the negative α axis. By starting the direct calculation of the forward MKP hysteresis model from $T = \alpha_p$ in decrement of ΔT until $H(\alpha_{m-1}) \le Y_d$, it can be determined that $T_d \in [\alpha_{m-1}, \alpha_m)$. In this case, the final link of the interface to be created is a vertical one and locates on the column m. Suppose on the column m, there are

$e+1$ operators with the previous output states of +1, which correspond to the grid points $s_{mm} \sim s_{m,m+e}$, f operators with the intermediate output states previously, which correspond to the grid points $s_{m,m+e+1} \sim s_{m,m+e+f}$, and the rest with the previous output states of 0, which correspond to the grid points $s_{m,m+e+f+1} \sim s_{mK}$. Similar arguments as for $Y_d > Y_c$ can reveal that there are $f+1$ possible solutions for T_d. According to the expression for Y_d, i.e.

$$Y_d = \sum_{i=1}^{m-1}\sum_{j=i}^{K} \xi_{s_{ij}} \mu_{s_{ij}} + \sum_{j=m}^{m+e+n} \frac{T_d - \alpha_{m-1}}{\Delta T} \cdot \mu_{s_{mj}} + \sum_{j=m+e+n+1}^{m+e+f} \xi_{s_{mj}} \mu_{s_{mj}} \tag{11}$$

The formula for T_d in the case of decreasing output can be obtained as

$$T_d = \alpha_{m-1} + \frac{\Delta T}{\sum_{j=m}^{m+e+n} \mu_{s_{mj}}} \cdot (Y_d - \sum_{i=1}^{m-1}\sum_{j=i}^{K} \xi_{s_{ij}} \mu_{s_{ij}} - \sum_{j=m+e+n+1}^{m+e+f} \xi_{s_{mj}} \mu_{s_{mj}}) \tag{12}$$

where n denotes the number of the operators on the column m that are switched from the intermediate output states to the new lower ones. By replacing n with the value of $0,1,\cdots,f$ in turn in the above equation, all the possible solutions (i.e. $T_{d_1}, T_{d_2}, \cdots, T_{d_{f+1}}$) for T_d can be calculated and the exact value of T_d is determined as the maximum one of them, namely

$$T_d = H^{-1}(Y_d) = \max\left\langle T_{d_1}, T_{d_2}, \cdots, T_{d_{f-e+1}} \right\rangle \tag{13}$$

In the case of the target output is the same as the current one, i.e. $Y_d = Y_c$, the target input keeps the same value of the current input, i.e.

$$T_d = H^{-1}(Y_d) = H^{-1}(Y_c) = T_c \tag{14}$$

Using the equations (8)~(14) for the inverse MKP model, the input required for the hysteresis system to achieve a desired output can be predicted.

5. Numerical simulations

So far, the MKP model and its inverse model have been completely presented. Following that, they were applied to simulate the actual response of an SMA hysteresis plant, in order to evaluate their performances.

5.1 Matlab/Simulink programme

Based on the numerical formulas derived in the previous sections, the MKP hysteresis model and the inverse MKP hysteresis model were programmed as several functions in the computer language C, and then transferred to the S-functions compatible with Matlab/ Simulink by the use of its S-Function Builder. Referring to the corresponding S functions, two user-defined Simulink blocks, respectively named as sf_MKP and sf_Inverse_MKP were built to implement the MKP hysteresis model and the inverse MKP hysteresis model respectively in this software environment. The different simulation tasks were realized by using these blocks and the convenient post-processing was also available to display the results graphically.

5.2 Data preparation

Experimental work were carried out on a simple SMA hysteresis plant at room temperature, which was a commercial NiTi tension spring (Mondo-Tronics Inc., Canada) under a dead mass of 6N. The specifications of this NiTi tension spring is shown in Table 1. In the experiment, it experienced one complete actuation cycle and a number of partial actuation cycles by alternately passing an electric current of 2A to heat it and switching off the current to cool it. Its temperature and the corresponding displacement during this experimental process were measured by the use of a thermocouple and a LVDT respectively, and stored in the matrix format of T = [TimeValues DataValues] and Y = [TimeValues DataValues]. Such an experiment was repeated twice. Then two sets of data involving both major loop and minor loops (i.e. multiple loops) of this hysteresis plant were acquired. Prior to the simulation tests, they were filtered to remove the noise and disturbance.

Table 1. Specifications of the NiTi tension spring

Wire diameter (mm)	0.75
Coil diameter (mm)	6
Length (mm)	30
Nominal actuation temperature range upon heating (°C)	45 ~ 65

The discrete Preisach plane for the MKP model and the inverse MKP model was defined with the specific values of the related parameters as given in Table 2. Then the parameters of the models (i.e. the weighting values for the MKP operators) were identified by means of the simple least-squares fitting of one set of measured data, which was represented by $\hat{\mu}$. Fig.8 (a, b) illustrates the weighting values in the Preisach plane and in 3D curves respectively. According to the distribution of non-zero weighting values, it can be estimated that the actuation temperature ranges of the NiTi tension spring are about 30°C ~80°C upon heating and 25°C ~50°C upon cooling respectively. The other set of measured data were adopted as the input signal to the MKP model and inverse MKP model in the simulation tests.

Table 2. Definition of the discrete Preisach plane for MKP model and inverse MKP model

Discretization level	$K = 20$
Number of MKP operators	$N = \dfrac{K \times (K+1)}{2} = 210$
Input temperature range (°C)	$T \in [20,80]$
Nominal actuation temperature range Interval (°C)	$\Delta T = [80-20]/(K-1) = 3.16$

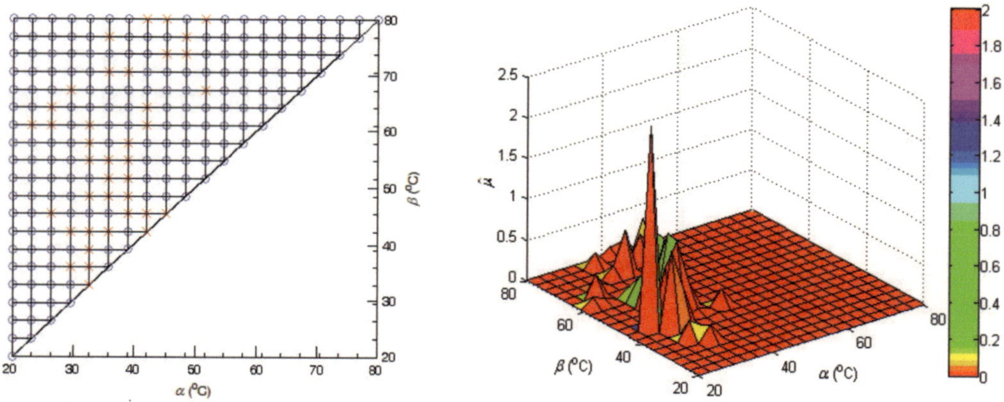

(a) Distribution of identified weighting values in Preisach plane (o: zero weighting value; ×: non-zero weighting value)

(b) 3D curve of identified weighting values

Fig. 8. Identified weighting values

5.3 Simulation tests

Two types of simulation tests were conducted here. One was the use of the MKP model to simulate the output of the NiTi tension spring in response to the measured temperature, while the other, on the contrary, was the use of inverse MKP model to predict the temperature, given the measured displacement. The graphical diagrams for these simulation tests by Matlab/Simulink are shown in Fig. 9.

(a) Simulation test on MKP model

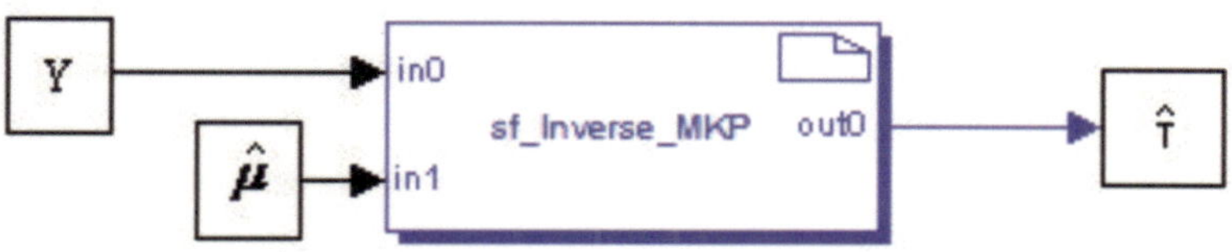

(b) Simulation test on inverse MKP model

Fig. 9. Matlab/Simulink diagram for the simulation tests

6. Results

6.1 Results of the simulation test on MKP model

Fig.10 illustrates the results of the simulation test on MKP model in terms of the modelled displacement vs. time curve, which is compared with the measured one. It is clear that the modelled displacement matches the measured one to a great extent.

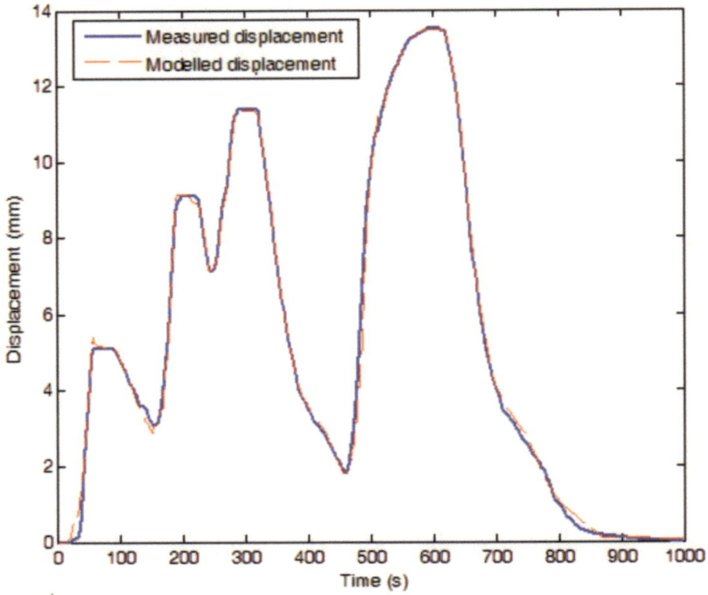

Fig. 10. Modelled displacement vs. time curve in comparison with measured one

The slight difference is mainly observed in the segments associated with the transition between a rising branch and a falling branch of the displacement curve. This phenomenon is more clearly demonstrated in the input temperature vs. output displacement curves as shown in Fig.11. The modelled multiple loops have a great resemblance to the measured one. To give a further quantitative analysis, the discrepancy between the modelled displacement and the measured one is calculated and plotted in Fig.12. Apparently, it is up to the bound of ±0.5mm, which is actually quite small.

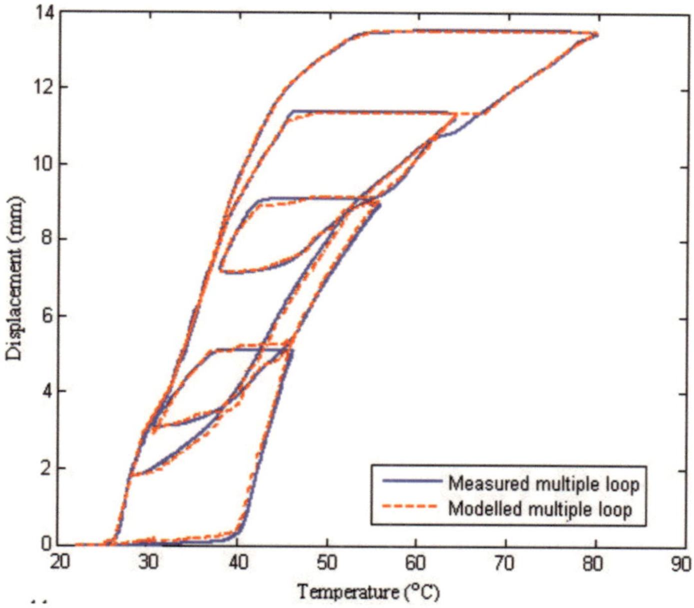

Fig. 11. Modelled displacement vs. temperature curve in comparison with measured one

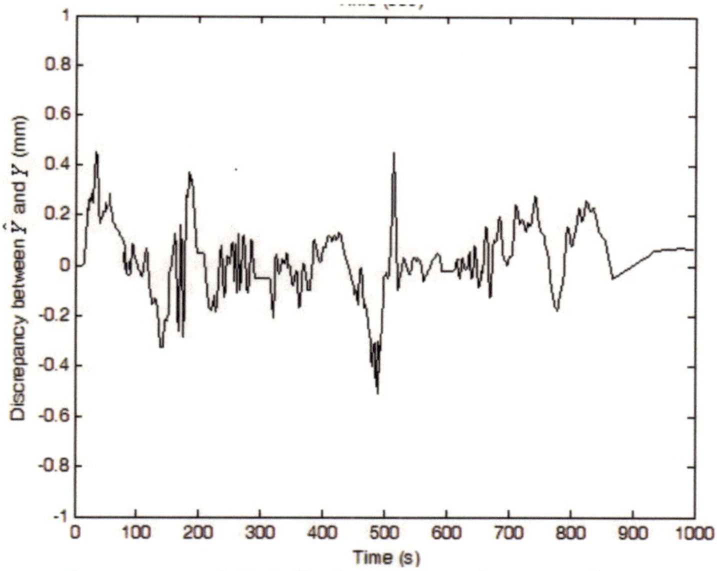

Fig. 12. Discrepancy between modelled displacement and measured one

6.2 Results of the simulation test on inverse MKP model

The temperature calculated from the inverse MKP model (i.e. the modelled temperature) in response to the measured displacement is shown in Fig.13 as a function of time. For comparison purpose, the measured temperature vs. time curve is also plotted here. It can be seen from the figure that the modelled temperature is very close to the measured one for the most of time. Exceptionally, when the measured temperature starts to rise or fall each time, the modelled temperature fails to follow it immediately. There seems to be a lag initially in each ascending or descending branch of the modelled temperature curve behind the measured one. The time history of the discrepancy between them is presented in Fig.14 , which reveals this phenomenon more clearly.

Actually, this phenomenon is resulted from the hystersis behaviour of SMA. Referring to the measured temperature vs. displacement curve (marked in bold line, Fig.11), each time when the measured temperature alternates the change direction, the measured displacement does not change with the measured temperature immediately. Instead, it keeps constant for some time while the measured temperature rises or falls. Moreover, recalling to the previous deduction process(see section 4), the inverse MKP model is actually deduced in such a way that its output (i.e. modelled temperature) changes when the input displacement changes, while keeps the value in response to the previous displacement, when the subsequent displacement is the same as the previous one. Take the fourth descending branch of the modelled temperature curve for example. To be able to view the details, close-up view of the modelled temperature, measured temperature and measured displacement curves from 550s to 700s is illustrated in Fig.15. Apparently, during the period of 586s~622s when the measured displacement, as the input to the inverse MKP model, is almost constant at about 13.5mm, the modelled temperature stays at a constant value of about 79°C, while the measured temperature keeps falling from 79°C to 53°C. As a result, the discrepancy between them increases rapidly and reaches the maximum value 26°C at the time instant of 622s (see Fig.14). Then at the next time instant (i.e. 623s), as the measured displacement starts to decrease, the modelled temperature drops to about 51°C and gradually catch the measured one since then. The discrepancy between them is less than 2°C, following the further deceasing measured displacement.

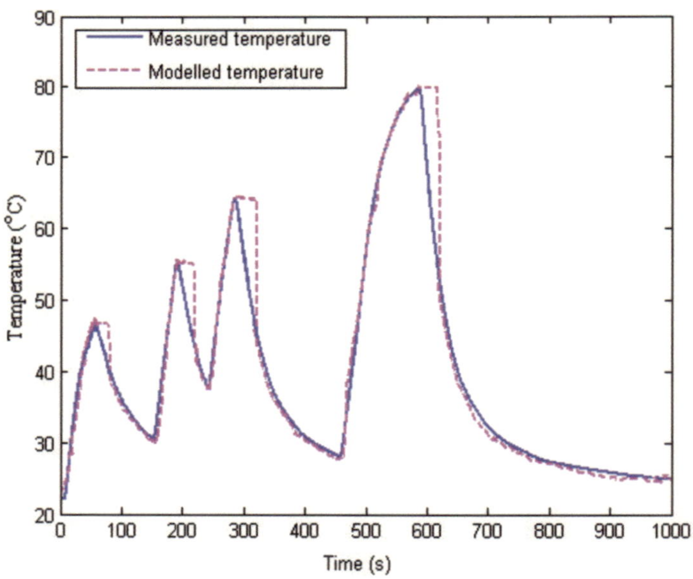

Fig. 13. Modelled temperature vs. time curve in comparison with measured one

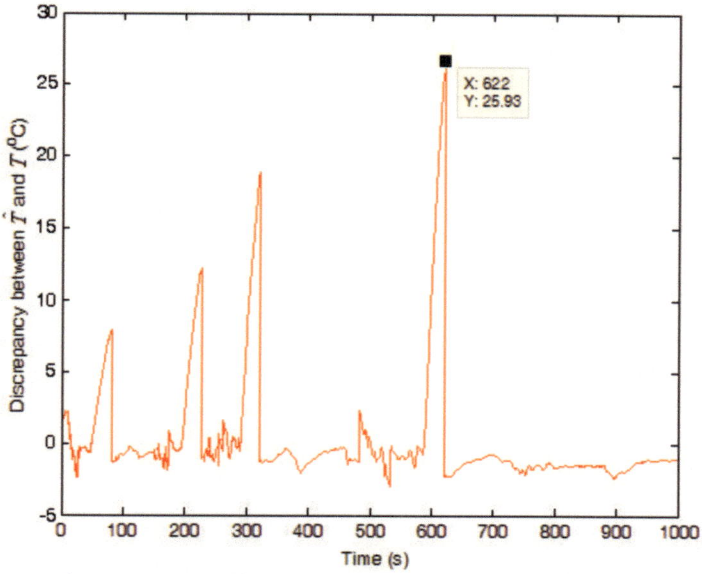

Fig. 14. Discrepancy between modelled temperature and measured one

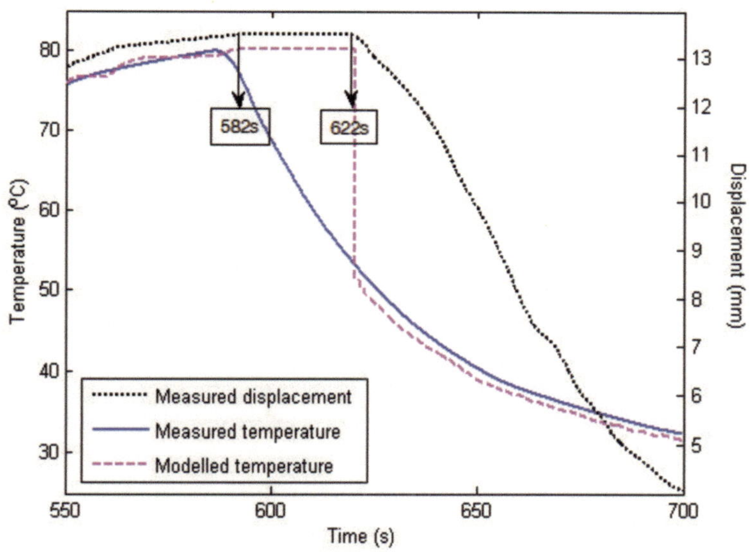

Fig. 15. Close-up view of measured displacement, measured temperature and modelled

7. Discussion

In the laboratory situation, there are always some degrees of measurement error in the data acquired mainly due to the accuracy of the sensors used in measuring the temperature of SMA actuators, e.g. thermocouple, and in measuring their displacement, e.g. LVDT. The accuracy specification for thermocouple and that for LVDT used here are ±1 °C and ±0.5 mm (i.e. ±1% of its measurement range of 50mm). Meanwhile, the weighting values identified by means of the least-square fitting are not the exact ones. These factors mainly

account for the small discrepancy between the modelled displacement and the measured displacement, and the one between the modelled temperature and the measured temperature when the measured displacement changes.
It is also important to point out that the accuracy of MKP model and inverse MKP model strongly depends on the parameter identification. These characteristic parameters of the models are directly connected with the macroscopic experimental facts. Typically, they are determined by means of simple interpolation of the limited experimental data or least-squares fitting of a measured major loop or multiple loops for a pre-selected load (Torrie & Vajida, 1994; Song et al, 2001; Ktena et al, 2001). Such identification strategies are strongly sensitive to measurement errors. More seriously, the hysteresis behaviour of SMA actuators varies with the applied stress, pre-deformation (see Fig. 2) and the number of thermal/ mechanical cycles. Therefore, such an off-line identified hysteresis model as an input-output static mapping cannot accommodate the dynamic changes and even introduces input error, when it is ill-matched to the real response of the hysteretic system. For example, if the loading condition of the NiTi tension spring were changed from 6N to 10N, the MKP model identified for it under a load of 6N here could not portray its hysteresis behaviour under a load of 10N. In the real control applications, SMA hysteresis plant is often associated with significant uncertainties such as the various loading conditions. Obviously, the off-line identified MKP model and inverse MKP model is unable to capture the real hysteris response. Instead, on-line update of the parameters should be incorporated in the control scheme based on MKP model and inverse MKP model, in order to achieve the accurate control.

8. Conclusions

MKP model proposed in this study is independent of the specific plant and thus is generally applicable to different hysteretic systems incorporating SMA actuators. The condition is that Preisach plane must be defined to cover the actuation temperature range of SMA actuators in use, which is rather weak and can be easily satisfied. Its advantage in the reduced computation is obvious and makes it more suitable for control purpose. The accuracy of MKP and inverse MKP models strongly depends on parameter identification. The results of the numerical simulation have successfully demonstrated that MKP and inverse MKP models are able to model and predict the response of SMA hysteresis plants, when their parameters identified by the use of the limited experimental data are sufficiently accurate. Since the off-line identified models cannot accomodate the unmodelled dynamics, it is necessary to correct them on-line to achieve hysteresis compensation for SMA hysteresic system with unknown dynamics such as various loading conditions in the real control applications. Noticeably, the expression of MKP model in the linearly parameterized form allows the general adaptive method, such as the gradient estimator, to be a handy choice for on-line parameter update. The findings are encouraging for moving to the development of control schemes based on MKP and inverse MKP models.

9. References

Bhattacharyya, A. & Lagoudas, D. (1997). Smart Materials &Structures, 6, 3, 235-250

De Sars, V.; Haliyo, S. & Szewczyk, J. (2010). Mechatronics, 20, 2, 251-264

Dickson Carrie, A.; Hughes Declan, C. & Wen John, T. (1996). Hysteresis in shape memory alloy actuators: the control issues, Proceedings of SPIE, 2715, 494-506

Gupta, K.; Darpe, A.; Kumar. B.; et al. (2009). Advances in Vibration Engineering, 8, 3, 247-254

Hughes, D. & Wen, J. (1997). Smart Materials & Structures, 6, 287-300

Krasonselskii, M. & Pokrovskii, A. (1989). Systems with Hysteresis, Springer-Verlag, Heidelberg

Krasonselskii, M.; Pokrovskii, A.; Rachinskii, D.; et al. (1994). Automation and Remote Control, 55, 7, 964-973

Ktena, A.; Fotiadis, D. ; Spanos, P.; et al. (2001). Physica B, 306, 84-90

Matsuzaki, Y. & Naito, H. (2004). Journal of Intelligent Material Systems and Structures, 15, 2, 141:155

Mayergoyz, I. (1991). Mathematical models of hysteresis, Springer-Verlag, Berlin

Okamura, H.; Yamaguchi, K. & Ono, R. (2009). International Journal of Optomechatronics, 3, 4, 277-288, 1559-9612

Popov, P. & Lagoudas, D. (2007). International Journal of Plasticity, 23, 4, 1679-1720.

Ŝittner, P.; Stalmans, R. & Tokuda, M. (2000). Smart Materials & Structures, 9, 452-465

Song, C.; Brandon, J. & Featherston C. (2001). Distributed-element Preisach model for hysteresis of shape memory alloys, Proc Instn Mech Engrs, 215 (part C), 673-682

Torrie, E. & and Vajda, F. (1994). IEEE Transactions on Magnetics, 30, 6, 4987-5000

Van Humbeeck, J. & Stalmans, R. (1998). Characteristics of shape memory alloys, In: Shape Memory Materials, Otsuka, K. &Wayman, C. (Ed.), 149-184, Cambridge University Press, Cambridge

Wang, J. & Dai, H. (2010). International Journal of Plasticity, 26, 4, 467-487

Wayman, C. & Duerig, T. (1990). An Introduction to martensite and shape memory, In: Engineering Aspects of Shape Memory Alloys, Duerig, T. (Ed.), 3-20, Butterworth - Heinemann Ltd., London

Webb, G. & Lagoudas, D. (1998). Journal of Intelligent Materials and Structures, 9, 432- 448

Wolfe, T.; FaulKner, M. & Wolfaardt, J. (2005). Smart Materials & Structures, 14, 4, 759-768, 0964-1726

Wong, F.; Rabbath, C.; Hamel, N.; et al. (2007). Transactions of the Canadian Society for Mechanical Engineering, 31, 1, 19-38

5

Experimental Study of a Shape Memory Alloy Actuation System for a Novel Prosthetic Hand

Konstantinos Andrianesis, Yannis Koveos,
George Nikolakopoulos and Anthony Tzes
University of Patras
Greece

1. Introduction

Recently, the development of compact, light-weight and powerful actuation systems has been in the centre of investigation at many scientific institutions and research groups all over the world. These systems can be used in devices of almost every aspect of modern life and based on their inherent technology they come with certain benefits and costs. One of the most demanding applications field in terms of actuator selection and design is the field of upper-extremity prosthetics. Modern commercial advanced hand and arm prostheses are conventionally actuated by electric servomotors. Although these motors achieve reasonable kinematic performance, they have been proven insufficient in meeting amputees' demands, mainly due to their noisy operation and limited energy density which leads to the use of bulky and heavy driving systems (Herr, 2003). Therefore, an alternative nonconventional actuation technology is requisite in order to overcome these limitations which make a substantial proportion of upper-limb amputees avoiding the use of their prostheses.

One of the most promising actuation technologies is based on Shape Memory Alloys (SMA) and phenomena related to change of their atomic structure. SMA are metallic alloys that can exhibit an actuation mechanism resembling the biological muscle - they contract producing actuation forces. These muscle-like actuators present high power to weight ratio enabling the development of compact, lightweight prosthetic devices without too much compromising power capabilities and eliminating the forced-tradeoff between dexterity and anthropomorphic size, weight and appearance (Bundhoo, 2009). Additional benefits include an inherent position feedback method (given a near linear relationship between ohmic resistance and contraction), silent, smooth and life-like operation, and the lack of requirement for force or motion transmission devices (Kyberd et al., 2001).

During the last two decades SMA have been studied and reviewed as possible actuation technology in prosthetics by many researchers but efficiency and response time are claimed as the most limiting factors (Del Cura et al., 2003). So, in order to render this material appropriate for application in upper-limb prostheses, these impediments must be overcome. Towards this scope, an innovative SMA actuation system for a newly developed prosthetic hand is constructed and studied. The technology applied in this hand offers a series of improvements when compared to current commercial prosthetic devices. Its design proposes the development of a multifunctional, silent and cosmetic appealing hand that will enable patients to carry out basic daily tasks more easily and reduce the rejection rate of prostheses.

In the next Section, a brief background of this research is provided. The development of the hand prototype along with its SMA actuation system will be analytically described in Section 3. In Section 4, the performance of this actuation system is assessed for a one-finger testbed while in the last section conclusions and future work are discussed.

2. Background

2.1 Prosthetic Hands

The human hand plays a vital role in every aspect of our daily life as it defines our interaction with the material world. Its functional uniqueness arises from the complex geometric arrangement of joints, ligaments and muscles enabling a massive array of movements and uses (manipulation, prehension and exploration) accounting for up to 26% of the human body's movement potential (Brown, 2008a). But because of this complexity, which is the result of evolution's continuous adaptation to the requirements of universal gripping and holding tasks over millions of years, human hand cannot be imitated expediently with the technological means available today. As a consequence, loss of such an important organ could be devastated.

Much work over the recent years have been involved in the creation of an anthropomorphic hand, where degrees of freedom and size of the hand and forearm are designed to be as close as possible to human proportions; some impressive robotic devices can be found in literature such as the NASA's Robonaut Hand, DLR Hand, Gifu Hand and Shadow Hand (Biagiotti, 2002). However, these devices fail to operate as prosthetic hands since this type of application imposes a series of challenging requirements (considerably different for example from the ones imposed in industrial robotics context, where the industrial hands typically operate in a structured environment with predefined tasks and, often, with no considerations to the human interaction). These requirements, as cited by many authors (Kyberd et al., 2001; Schultz et al., 2004) based on surveys about amputees' needs, can be summarized to the following: low weight, noiseless operation, anthropomorphic size and appearance, high functionality and dexterity, compactness, easy controllability, low cost, user safety and reliability.

The majority of users has one good hand (unilateral amputees), and become quite adept at performing tasks one-handedly, with little involvement of the prosthetic hand. So, they need the prosthesis merely as an assistor to their normal arm, and do not want to be bothered with complex devices which need care and maintenance and great concentration and practice in order to gain proficiency. Besides, if the prosthesis does not make life any easier for the user, it is likely to be rejected.

There are a wide variety of prosthetic devices available for upper limb amputees ranging from purely cosmetic ones to advanced myoelectrically controlled bionic hands. The reader can address to (Lake and Dodson, 2006; Trost and Rowe, 1992; Kelly et al., 2009) for an in-depth review of pros and cons of the different types of upper-limb prostheses. Current commercial high-tech prostheses use small electromagnetic motors installed into a human-like terminal device and are typically controlled by one or two surface electromyography (EMG) electrodes that detect muscle contraction signals in the remnant limb of the user. At present, the best known commercially available prosthetic hand is the i-Limb™ Hand (Fig. 1a). It was first launched in 2007 by Touch Bionics and until now more than 1000 patients all over the world have fitted with it. It has 5 independent digits each controlled by a small conventional electric motor along with a combination of planetary drives, worm gears and cables. The fingers have three phalanges, like the human hand, with a stationary end joint. The knuckle joint is motor driven and motion transmitted to the middle joint by a tooth belt. Hence, a single motor can cause the whole finger to bend. This approach of having less actuators than degrees of freedom (DOF) is said to be an underactuated approach providing compliance to the fingers. The thumb is also controlled by a motor and is able to rotate so that it closes to meet the fingers.

Before i-Limb, amputees' choice was limited in prosthetic hands that behave as simple grippers with one DOF. In these devices, a single motor drives the first and second fingers of the hand in unison with the thumb to produce a tri-digital grip (open or close); the two fingers are actually one rigid link, in opposition to a rigid thumb. The remaining two fingers on the hand are for aesthetic purposes and move passively with the first two fingers. Some examples of these prostheses include the Otto Bock's SensorHand™ Speed (Fig. 1b) and the Motion Control Hand (Fig. 1c). These electric prostheses do not allow adequate encirclement of objects, in comparison with the adaptive fingers of i-Limb Hand, resulting in unstable grasps in presence of external perturbations.

Taking advantage of the significant technical improvements of the last years in various fields, the development of better and more functional prostheses is feasible. Progress in materials science, microcontrollers, batteries and cosmetics products permit the design of advanced prosthetic devices able to help amputees' life to be minimal affected.

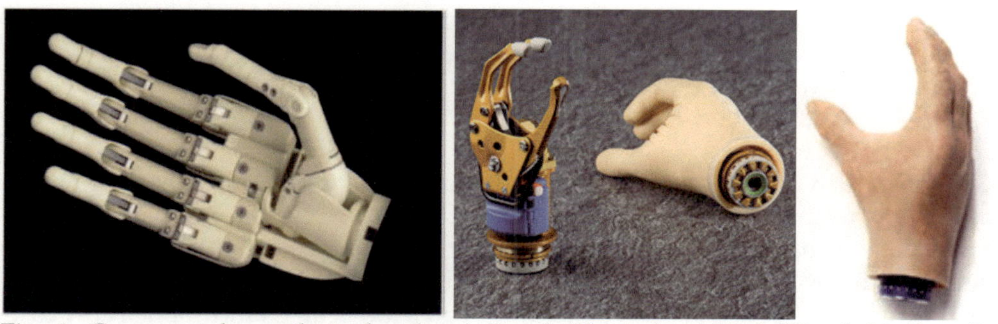

Fig. 1. Commercial prosthetic hands: a) Touch Bionics's i-Limb Hand, b) Otto Bock's SensorHand Speed, c) Motion Control Hand

2.2 Shape Memory Alloys

Shape memory alloy (SMA) is a smart material that has the ability to return to its predetermined dimensional configuration when heated beyond a threshold phase transformation temperature (Mavroidis et al., 2000). This behavior occurs due to a change in the material's crystallographic structure between two phases: the low temperature (martensite) and high temperature (austenite) phases (Fig. 2). Austenite and martensite are identical in chemical composition, but have different crystallographic structures. When an SMA is deformed in martensite, the residual strain can be recovered by heating the material to the austenite phase. This Shape Memory Effect (SME) returns the SMA to its original shape. Thus this property can be exploited for the design of devices/actuators. This is done by training the material to remember a specific shape in the austenitic condition. When used as linear actuators, SMA are commercially available as pre-strained (trained to remember a shorter than actual length on heating) martensitic wires. Joule heating beyond the transition temperature triggers the phase transformation to austenite where the stretched wire contracts to the pre-strained length. During this transformation, the SMA can yield an extremely large force if it encounters any external resistance. Now, when the contracted wire is cooled, it returns to the martensite state (twinned) where the material is malleable. A reverse-bias force is needed to return the wire to its original length. Bias forces can be created by many methods: gravitational pull, spring, magnetic force, opposing SMA wire.

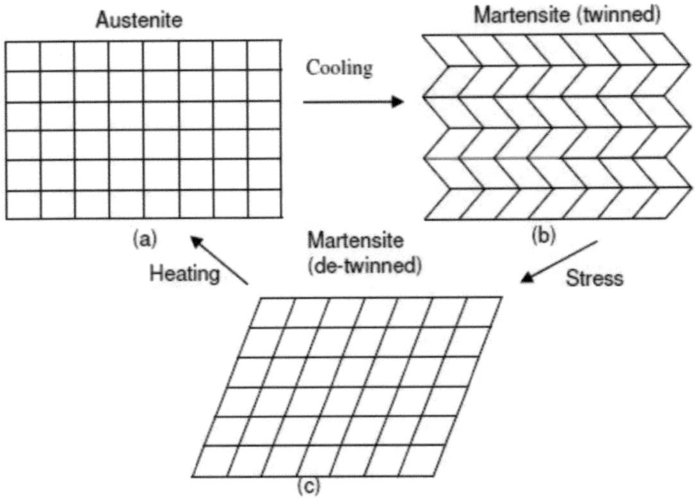

Fig. 2. Grid-like representation of SMA structure

The SME described above, whereby only the parent austenitic phase is remembered by the alloy, is referred to as the one-way SME. It is however possible to make an alloy remember both the parent austenite phase and the martensite shape simultaneously. This is referred to as the two-way SME. In this case, the alloy exhibits two stable phases: a high-temperature austenite phase, apparent on heating and a low-temperature martensite phase, apparent on cooling. Although SMA of two-way SME provides contractive and tensile forces, its tensile force is much smaller than contraction force and recoverable strain normally less than half that of one-way type (Lan et al., 2009). Thus SMA actuators of one-way effect are more attractive in robotic applications and usually preferred.

Next, a brief review of the advantages and limitations of SMA for actuation purposes is presented.

Advantages

- Compact, lightweight with high power/mass ratio and energy density - Comparing a large selection of actuation technologies, SMA actuators feature the highest power to mass ratio at less than 0.45kg masses. The energy (work) density of SMA is also very high; between 5000 to 25000 KJ/m^3 when human muscles in comparison exhibit between 40-70 kJ/m^3 (Abolfathi, 2007).

- Ease of actuation and low voltage requirement - Various methods can cause thermal activation on the SMA but since these are inter-metallic alloys, they can be easily driven by electrical current via Joule heating. Also, the small voltages that are required to operate make them safe for human-oriented applications.
- Clean, silent and spark free operation - In contrast to many actuators such as electric motors, SMA actuators operate with no friction or vibration allowing extremely silent movements; this is a strong asset for the prosthetic applications where the conventional technology used remains quite annoying and uncomfortable for the users. They are also free of parts such as reduction gears and do not produce dust particles.
- Long actuation life - When used within sustainable strain and stress limits, SMA actuators can be expected to last hundreds of thousands of operation cycles.
- High biocompatibility and excellent corrosion resistance - This enables their use in an environment with high humidity or wet.
- Direct-driven - Using SMA as actuators there is no need for complex and bulky transmission systems. This drastically reduces the complexity of the hand's driving mechanism.

Limitations

- Low displacement levels - Even though SMA exhibit relatively large strains, only a fraction of the net strain can be utilized in order to maximize the actuator lifetime. So, long lengths of SMA wire are required for large strains. However, using appropriate arrangements that convert the small strains into large motions this limitation can be moderated.
- Low power efficiency - SMA operate through heat and as such are limited by the Carnot efficiency model to at most 10%.
- Low operating frequency - The rate at which an SMA actuator can shift to austenite phase and return to martensite phase is limited by the slow heat transfer processes (low thermal conductivity) needed to promote the phase transformation of SMA. Typically, operation is faster for actuation (heating) than it is for relaxation (cooling). So, mostly the level of cycles per minute is dependent upon the rate of cooling of the wire. A variety of methods have been proposed to increase cooling speeds such as water immersion, heat sinking and forced air. However, even if these methods improve the bandwidth, they also cause an increase in power consumption as more heat is required to actuate the wire within the cooling medium.
- Control difficulties - Hysteresis, nonlinearities, parameter uncertainties and un-modeled dynamics introduce difficulties in accurate control of SMA. The SME is not a thermodynamically reversible process. Heat losses during the phase transformation phases (owing to internal friction or structural defects) cause hysteretic behavior of SMA as shown in Figure 3.

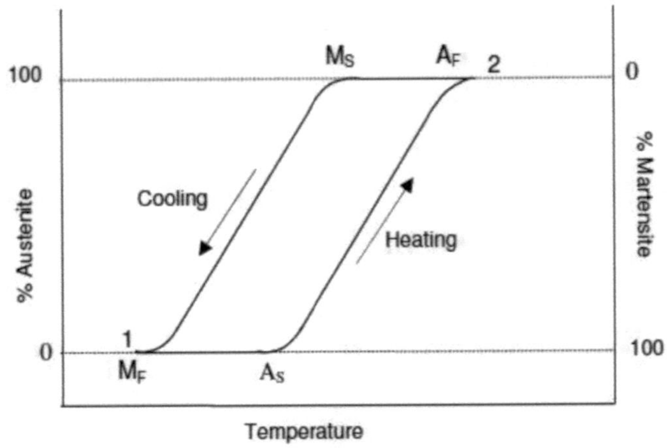

Fig. 3. Hysteresis loop in SMA

Out of all the SMA that have been discovered so far, Nickel-Titanium (NiTi) -well known as "Nitinol" (which stands for Nickel (Ni), Titanium (Ti) and US Naval Ordinance Laboratory (NOL) where the alloy was discovered in the early 1960s by William Beuhler) has proven to be the most flexible and beneficial in engineering applications showing the best combination of properties. The following characteristics of Nitinol make it stand out from the other SMA: greater ductility, more recoverable motion, excellent corrosion resistance, stable transformation temperatures, high biocompatibility and low manufacturing costs (Teh, 2008).

2.3 Related Research

Considerable efforts have already been made from different research groups worldwide to develop articulated robotic hands or finger mechanisms using SMA as the power elements of their actuation system. More specifically, DeLaurentis and Mavroidis developed an aluminum finger with 4 DOF, using bundles of Nitinol wire strands 150 microns in diameter as artificial muscles to move its joints (Fig. 4a). As a continuance of this work, an SMA actuated hand (the Rutgers Hand) was rapid prototyped comprising of 5 fingers and 20 DOF (DeLaurentis and Mavroidis, 2002). Price et al., 2007 showed practical application of shape memory alloys as actuators (Price et al., 2007) in a three-fingered robotic hand (Fig. 4b). Maeno and Hino proposed a miniature five fingered robotic hand for dexterous manipulation of small tissues and parts in medical and industrial fields, driven by SMA wire actuators with diameter of 0.05 mm (Maeno and Hino, 2006). The size of this hand was about one third of the human hands and it had 4 DOF per finger and 20 DOF in total (Fig. 4c). Cho et al. presented a five-fingered robotic hand with 16 controlled DOF and 32 independent SMA axes demonstrating the advantages of joule-heated Segmented Binary Control where each SMA actuator wire is divided into several segments (Fig. 4d). The total weight of this hand system was about 800 grams (Cho et al., 2006b). O'Toole and McGrath with their work (O'Toole and McGrath, 2007) also proposed a prosthetic hand design that incorporates embedded SMA bundle actuators. Bundhoo et al. constructed a 4-DOF (three active and one passive) artificial finger testbed (Fig. 4e) based on the combination of compliant tendon cables and one-way SMA actuators (off-the-shelf products from Miga

Motor Company) in an antagonistic arrangement for the required flexion/extension or abduction/adduction of the finger joints (Bundhoo et al., 2008).

Besides the above research efforts that took place in the last decade, it is worth referring one of the very first attempts to introduce SMA in robotic hands: the Hitachi Hand. This hand caused an understandable sensation with its debut in 1984 claiming a 10:1 reduction in weight as compared to other hand designs. It had three 4-DOF fingers and a thumb. It also possessed a forearm and a pitch-yaw wrist. It used a large number of SMA wires, actuated in parallel. Electrical heating of the wires resulted in their contraction against a force spring. On cooling, the wires returned to their original length. The force generated on contraction was used for joint actuation. The Hitachi Hand used 0.02 mm diameter SMA actuators for the fingers. Each DOF of the wrist, on the other hand, were actuated by 0.035 mm diameter SMA wires, set around pulleys. The above mechanism enabled 90° joint travel. Joint positions were sensed by potentiometers. This four-fingered gripper was 69.85cm long, weighed 4.49kg and was capable of holding a 2kg load capacity (Yang and Wang, 2008).

However, in each of the above cases few experimental results are available, and no complete feasible solution has been proposed to satisfy the requirements of a prosthetic device.

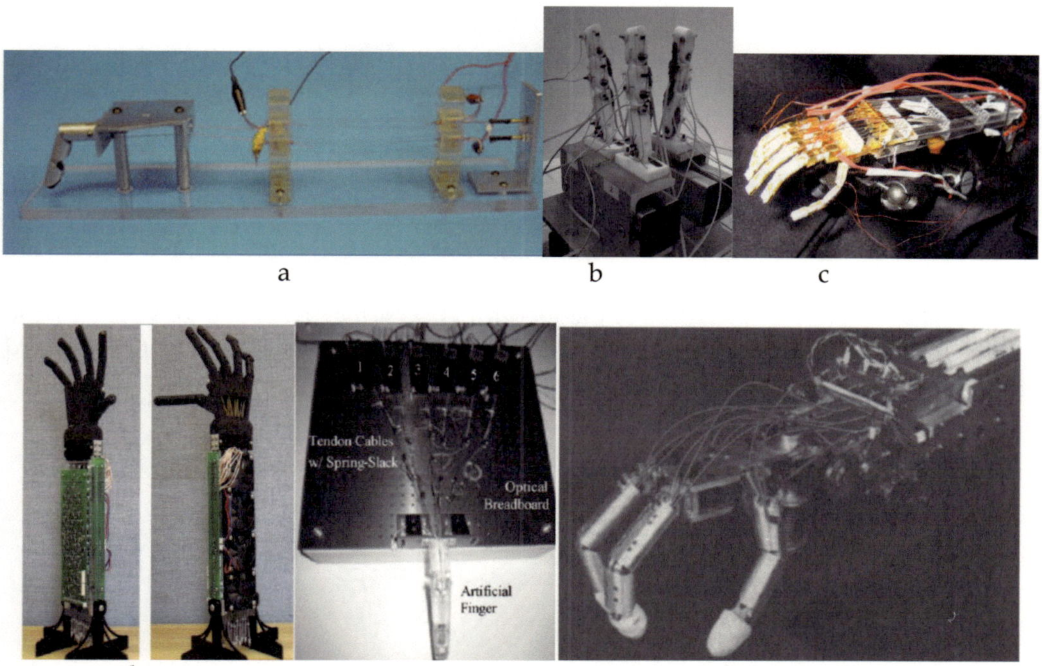

Fig. 4. Various SMA actuated mechanisms

3. Prosthetic Hand Development

3.1 Mechanical Design

Human hand is a biological system that has evolved into a very efficient and effective mechanism after millions years of evolution and as such it is the unchallengeable benchmark of our research efforts. Its advanced and complex kinematic format gives dexterity, combined with delicacy of movement and capacity for power grasping, and endows humans with incredible powers of maneuverability and capability for interaction through various object grasping and manipulation strategies. At the same time the hand and fingers also have a sensory function fundamental to these exploring purposes. Thus, an extensive study of natural hand biomechanics and anatomical data stands to reason when designing a new prosthetic device.

The human hand is a very articulated structure and the most dexterous part of the human body. It consists of 27 bones, 8 of which are located in the wrist. The other 19 constitute the palm and fingers as shown in Figure 5. The bones in the skeleton form a system of rigid bodies connected together by joints with one or more DOF for rotation. Joints between the bones are named according to their location on the hand as metacarpophalangeal (MCP) (i.e., joining fingers to the palm), interphalangeal (IP - the proximal interphalangeal (PIP) and the distal interphalangeal (DIP) joints) (i.e., joining finger segments) and carpometacarpal (CMC) (i.e., connecting the metacarpal bones to the wrist). The nine IP joints can be accurately described as having only one DOF, flexion/extension. All five MCP joints, however, are described in the literature as saddle joints with two DOF: abduction/ adduction (i.e., spreading fingers apart) in the plane defined by the palm, and flexion/extension. The CMC of the index and middle fingers are static while the CMC of the pinky and the ring finger have limited motion capability reflecting palm folding or curving, which is often discarded yielding a rigid palm. The CMC of the thumb, which is also called trapeziometacarpal (TM), is the most difficult to model. Biomechanical studies have shown that the TM joint has two non-orthogonal and non-intersecting rotation axes. The two DOF saddle joint is a restrictive model but it has been used in many studies (Albrecht et al., 2003). The structural limits of these joints are dictated by the anatomy and physiological make-up of the hand. Though the hand has a high number of freedoms, some of these degrees are interdependent owing to the muscular inter-connections of the hand. Fingers motion is performed by contracting muscles in the forearm and palm which are attached to the finger bones through a convoluted network of interdependently acting tendons and ligaments.

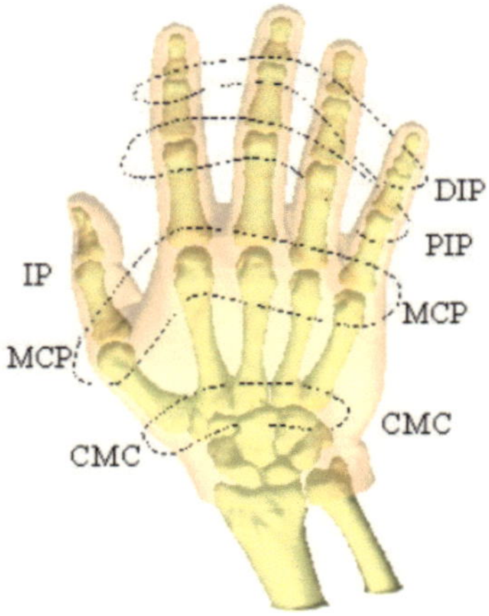

Fig. 5. Human hand joints

So, inspired by the nature's elegant and effective solution to the problem of actuation and exploiting nonconventional biomimetic forms of actuation such as SMA we proceed with the development of a novel prosthetic device that mimics human hand in appearance, size and performance as much as possible. Many design approaches were evaluated experimentally or via intense graphically simulations and eventually an optimized CAD model was created, illustrated in Figure 6. Its geometry is based on anthropometric measurements and is equivalent to the hand size of the average percentile American female (almost equal to the 5th percentile American male). It is designed along with a small forearm stump of 15cm long needed to accommodate all its necessary components such as its actuators and electronics. In this way, a very compact design is achieved allowing for the fitting of a high number of upper limb-deficient persons.

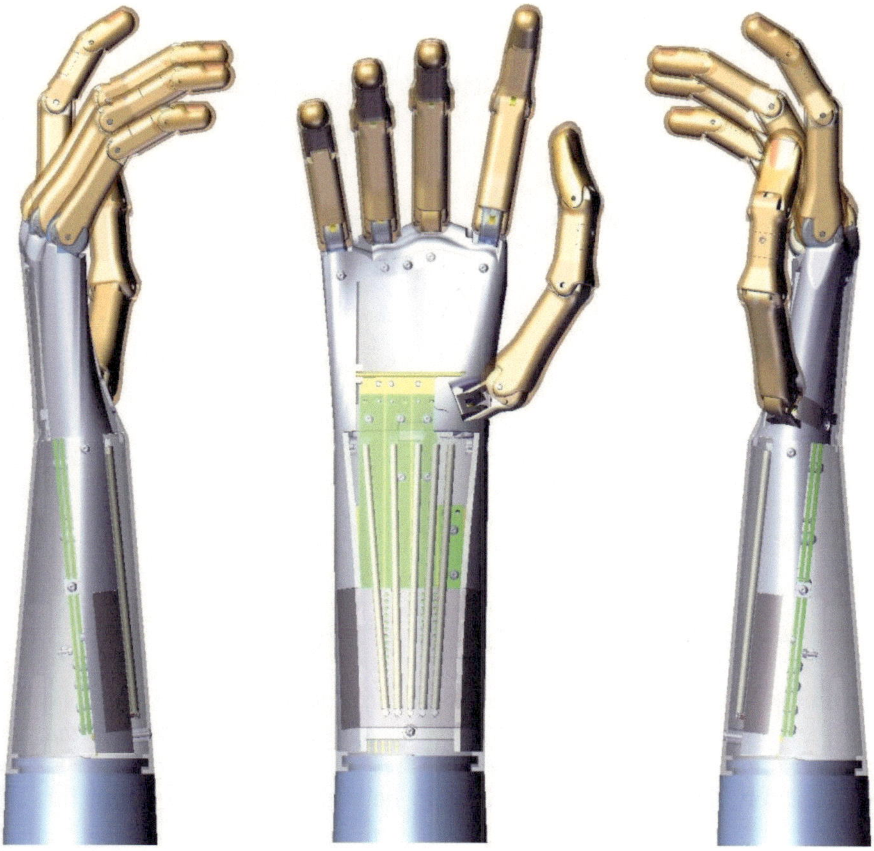

Fig. 6. CAD model of the prosthetic hand

It consists of five digits: a thumb and four fingers. In order to enhance its dexterity, a high number of DOF is requisite. Towards this goal, SMA due to their compactness can outmatch the traditional electro-magnetic actuators. However, since in myoelectric prosthetic hands only few control signals are available from the EMG control interface, it is not possible for the amputee to control many actuators (Carrozza et al., 2003). Therefore, an underactuated approach is adopted whereby a large number of DOF is controlled with a limited number of actuators. In this way, fingers can conform to the shape of an object during grasping (adaptive grasp). The geometric configuration of each finger and the orientation of each phalange are automatically determined by the external constraints imposed by the object, so that active coordination of the phalanges is not needed. The dimensions of the links, the configuration of the fingers, and the position of the contact points will determine the distribution of forces between the phalanges and the grasped object. In general, low contact forces are needed for grasping (below 10N for most manipulations) and contact forces are distributed similarly to those of natural hands.

The kinematic architecture is shown in Figure 7. These 19 DOF are controlled via 8 SMA Actuation Units (AU). More specifically, index and middle fingers utilize two AU per finger while one AU is granted to ring and pinkie. The last two are used for the actuation of thumb. Thus, all digits can move independently from each other. Finally, an extra DOF concerning forearm rotation is independently actuated via a high torque servomotor installed in the back of the forearm housing. As it is claimed by (Almstrom et al., 1981) this movement is considered to be the most useful after the grip for amputees.

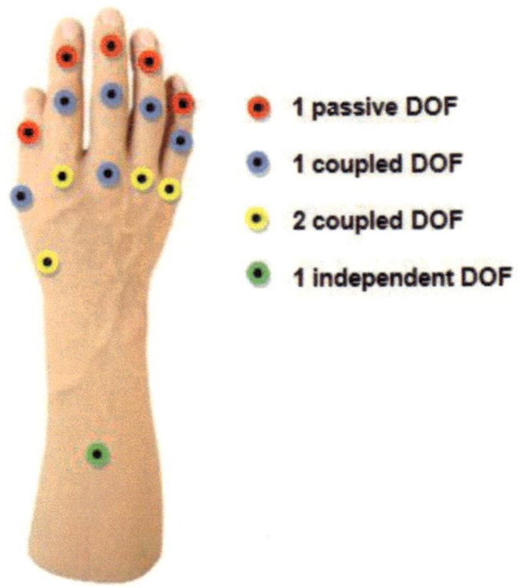

Fig. 7. Kinematic architecture of the proposed prosthesis

With this architecture, the formation of all the basic types of hand grasping as described by Schlesinger in 1919 and shown in Figure 8 is permitted. As it has been stated, 90% of all common dexterous actions are covered with these six basic grasp primitives. Thus, it is expected common activities of daily living to be easily performed using this device by amputees.

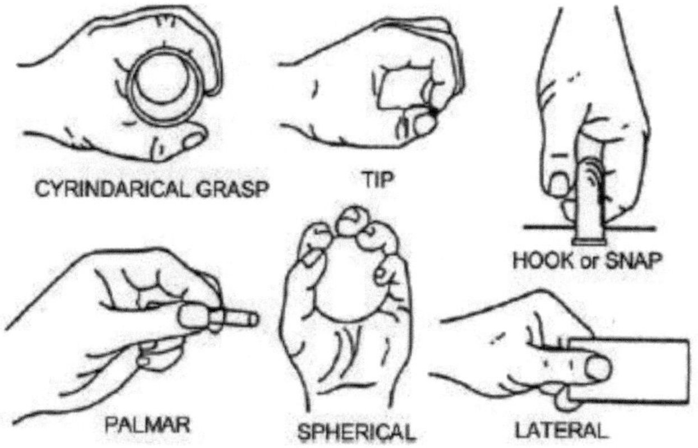

Fig. 8. Basic grasp primitives according to Schlesinger

3.2 Actuation Mechanism

SMA actuators are the key elements for the development of a lightweight, noiseless and multifunctional prosthesis. In this research, specially processed Nitinol wires, commercially available under the trademark "Flexinol" by Dynalloy Corporation, are used (Brown, 2008b). These wires are one-way actuators that when heated contract about 3-4% in length exerting significant stress (~200MPa). Their contraction-heating can be easily accomplished using an electrical current in less than one second. On the contrary, the cooling phase is much slower and a bias force is needed in order to have the wire returned to its relaxed uncontracted state.

Flexinol wires are available in various diameter sizes (ranging from 0.25 to 0.375 mm). Smaller diameter results in quicker cooling time but smaller force output. In fact, there is a trade-off between speed and force; while cooling time has an inversely proportional relationship with cross-sectional diameter, force output is inversely proportional to its square. There are also other factors that affect the cooling time of the SMA wire, such as the method of heating, the medium surrounding the wire while also there are methods that can increase force output such as bundling parallel wires or force-multiplying arrangements.

After intensive experimentation with these wires and different setups, we develop modular and compact AU, illustrated in Figure 9a, that have a high force output and fast time responses (heating and cooling times are discussed analytically in Section 4). For each AU, one Flexinol wire of 0.25mm diameter is used; based on the supplier this wire can last for more than 1,000,000 life cycles, exert forces up to 9N and needs 3.5 seconds for cooling. Bending this wire to the middle doubles the provided force to 18N (another way to augment the force output of each AU concerns the bundling of several parallel wires; however, it is proven in practice that implementation of this method encounters many difficulties). Also, given the limited strain capabilities during contraction and prosthetic hand's tight dimensions a special pattern for the Flexinol wire has been applied to all the AU (Fig. 9b). More specifically, the wire is wrapped around pulleys of 5.5mm diameter permitting an about 10% contraction in regards to AU's length. Moreover, in order to increase actuation bandwidth, Flexinol wires run through a small diameter rubber tube which acts as a heat sink helping them cool faster.

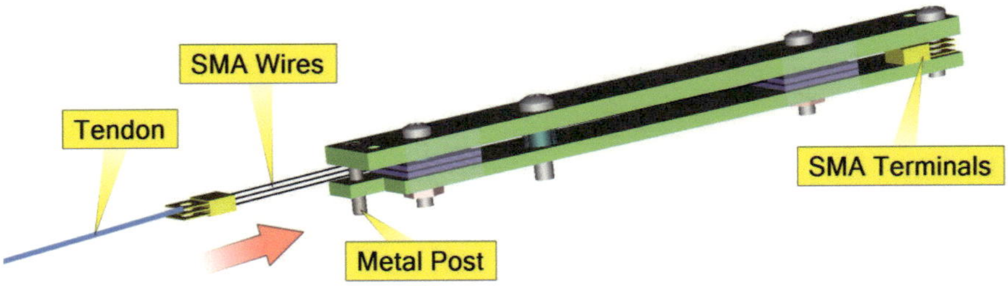

(a)

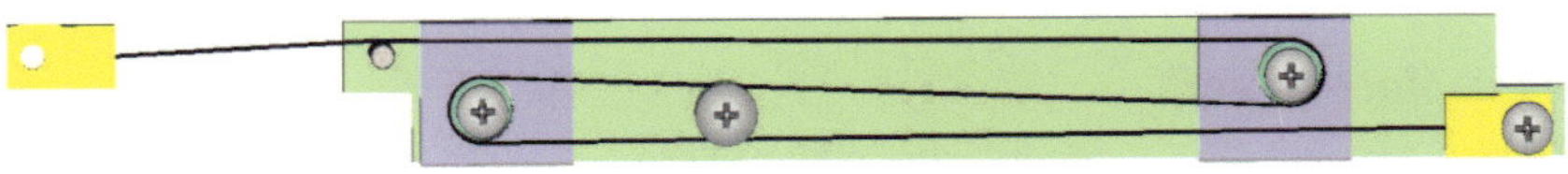

(b)

Fig. 9. An actuation unit and its a) working principle, b) SMA wire arrangement

As it is already mentioned, an underactuation mechanism is chosen allowing the digits to move independently. Their motion is controlled by 8 AU in total located intrinsically of the hand structure as shown in Figure 10 and rested on a printed circuit board (PCB) that brings all the necessary wirings.

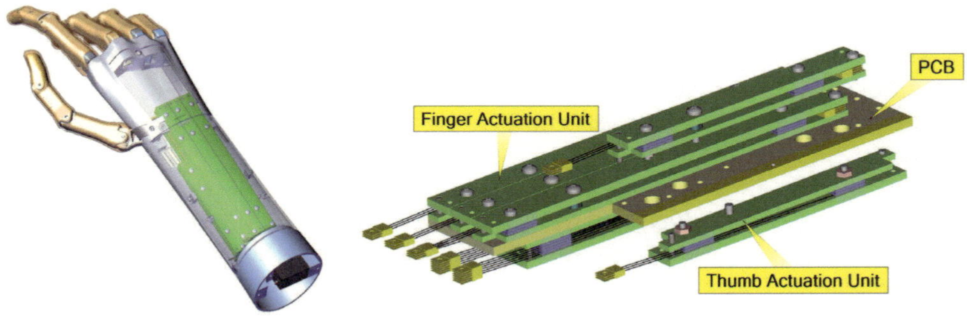

Fig. 10. Integration of an SMA actuation system in the structure of the prosthetic hand

In order to configure each AU's assignment for optimizing hand function, the following must be taken into account. Since SMA actuators have a very low power efficiency (<10%) and active hand prostheses are typically powered by rechargeable batteries of limited capacity, there is a need to reduce the power requirements of the hand as much as possible. Towards this goal, a voluntary opening mechanism is applied to our prosthesis. This means that the device is closed at rest and electrical power is required to open the hand. Pre-tensioned extension springs act antagonistically to the AU (which are controlling the opening of the hand) and flex all the digits providing a passive adaptive grasp. Apart from

the pre-tensioned springs, in case that higher pinch forces are requisite for a grasp, two additional AU give extra flexion force to the index and middle finger. In the thumb, two AU are responsible for its motion: one for extension and the other for adduction; extension springs are used for passive flexion and abduction.
A tendon transmission system is used in order to transmit force in the fingers. The tension of the tendons generates a flexing torque around finger joints, by means of small pulleys, allowing hand motion. Whereas biological tendons are made up of dense connective tissue which is elastic, flexible, and strong, in our design nylon-coated stainless steel wires are used as mechanical counterparts. They have their one end terminated into the middle phalange of each digit and the other end attached to an AU or extension spring. This implies that all the distal (DIP) joints of the digits are not actively actuated; in fact they are passively flexed about 25 degrees by means of a contraction spring between the last two phalanges.

3.3 Prototype Fabrication

The overall hand design consists of many different CAD-generated solid parts assembled in a manner that forms the suggested prosthesis structure. Using these three-dimensional models and a Rapid Prototyping (RP) technique (SLS process), a hand prototype has been fabricated and assembled as shown in Figure 11. For this purpose, a fiber-reinforced plastic composite with high strength and stiffness is used.

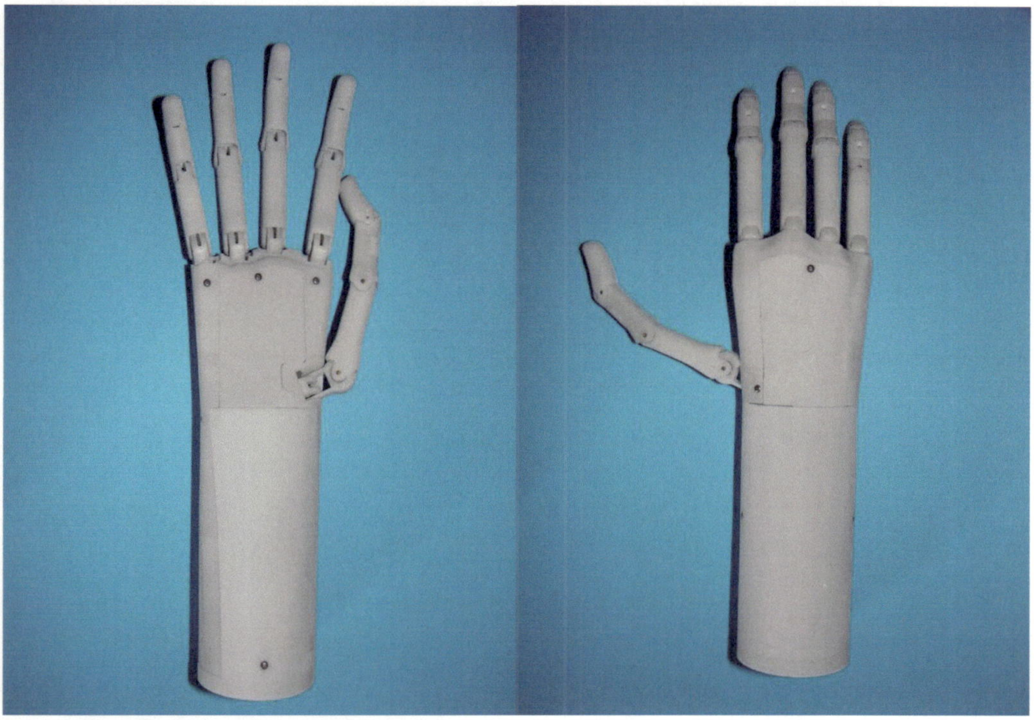

Fig. 11. Prosthetic hand prototype

Its weight is less than 200 grams and uniformly distributed along its length. With the integration of electronics and battery, it is estimated to weigh significantly less (about one third) than the current commercial prosthetic hands. Besides, as it is already mentioned, the mass of a prosthetic hand must be as low as possible in order to be accepted as a technical aid by the amputee; as it is worn on the end of a closely fitting external socket, its weight bears directly onto the skin of the stump. The lever-arm created is therefore large and the weight can obstruct blood flow in the underlying skin and results in symptoms ranging from discomfort to skin breakdown (Kyberd et al., 2001).

4. Experimental Study

4.1 Testbed

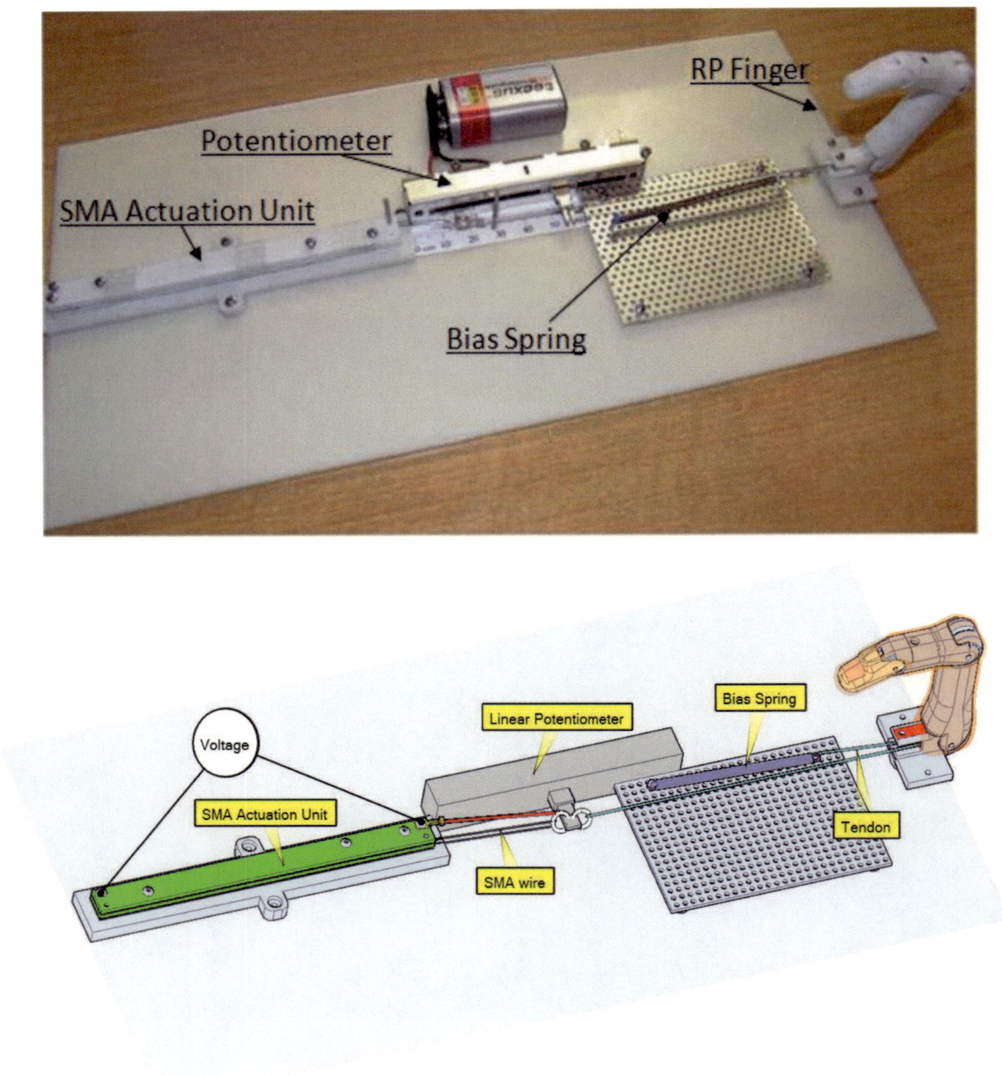

Fig. 12. Artificial Finger Testbed

As our prosthetic hand consists of five individual digits, the first phase of experimentation concerns a one-finger testbed developed as shown in Figure 12. It follows the aforementioned activation pattern: when AU is powered, the finger extends (opens) while an extension spring acts antagonistically to this motion in order to flex the finger back to its rest position. A linear potentiometer is used in order to measure the displacement generated by the AU. Additionally, miniature thermocouples are attached to the Flexinol wire of the AU for temperature monitoring purposes. A personal computer (PC) interfaces with the testbed through a PCI-6229 data acquisition card manufactured by National Instruments.

Actuation is controlled by electrical heating of the Flexinol wire. Though this can be done using a direct current (DC), Pulse Width Modulated (PWM) technique is preferred; PWM activation enables a more uniform heating of the SMA wires as compared to joule heating with a DC current (Bundhoo, 2009). Thus, it can improve the overall energy efficiency of our actuation system. As only a single voltage power supply is required, the need for additional voltage amplification equipment is eliminated. Moreover, PWM is easily implemented using microprocessors.

A low-cost development board (STM-H103) interfaced via a USB port on the PC and based on an ARM 32-bit Cortex™-M3 CPU is used for generating the PWM signals. The duty cycle (τ) and frequency of the PWM can be specified by writing to specific hardware registers. The PWM signals have a value ranging from 0 to 4095 (based on a 12-bit signal) resulting in PWM duty cycle ranging from 0% to 100% (Fig. 13). An external power supply provides the necessary voltage (Vo) and custom-made electronics are used to feed Flexinol wire with the desired current. Supervision and experimentation with this testbed is carried out through a graphical LabVIEW environment.

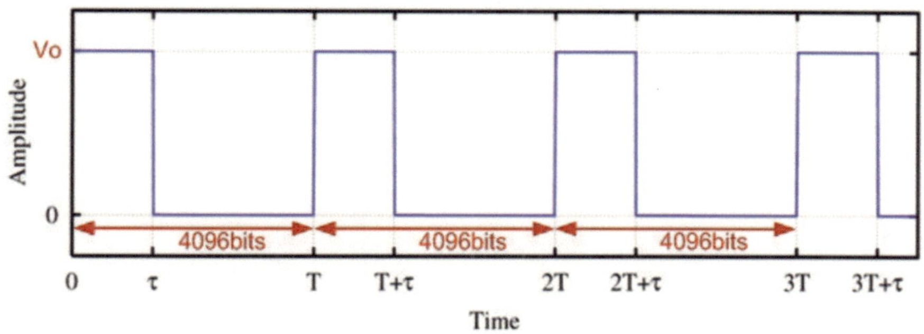

Fig. 13. A PWM output signal waveform

4.2 Results & Discussion

Open-loop Experiments:

As stated previously, the AU imparts motion to the finger via a tendon transmission system; a flexing torque is generated around the finger joints by means of 3mm radius pulleys. The range of motion for the controlled joints of our finger is: 90 and 75 degrees for the MCP and PIP joint respectively. So, the contraction length needed for extending fully the finger can be approximated by the following equation:

$$\Delta l_i = \sum_i R_i \theta_i = R_{MCP}\theta_{MCP} + R_{PIP}\theta_{PIP} = 3 \cdot \frac{90\pi}{180} + 3 \cdot \frac{75\pi}{180} = 8.64mm$$

The AU has greater strain capabilities than the ones needed. It can provide a 12mm displacement which is about 10% of its overall length. This means that the Flexinol wire will have quicker responses since it undergoes only a partial phase transformation in order to bring the finger in an open position.

Using the setup described above, experimental studies of open-loop control are conducted. During each experimental run, the following data is recorded: SMA voltage drop, current, contraction and temperature at the surface of the SMA. A third-order Butterworth filter with a 20 Hz cutoff frequency was used to smooth noise in the current data.

Towards evaluating finger's response, different electrical currents are applied to the SMA wire for an initial load of 5.32N (generated by a pre-tensioned extension spring with a constant k=0.28N/mm); when the displacement-contraction level measured by the linear potentiometer equals to 8mm (i.e. finger in open position), power is off letting unhindered relaxation-cooling of AU flexing back the finger. The resulting plots are depicted in Figures 14-17. From their examination, one sees that an increase of the amperage has drastic reduction in time needed for a full actuation cycle. Comparing the temperature curves (Fig. 15-16), a much higher temperature peak is observed for the lower current input. However, this higher temperature did not result in increased contraction as better shown in Figure 17. In other words, a significant amount of thermal energy was wasted since it was not converted into mechanical work.

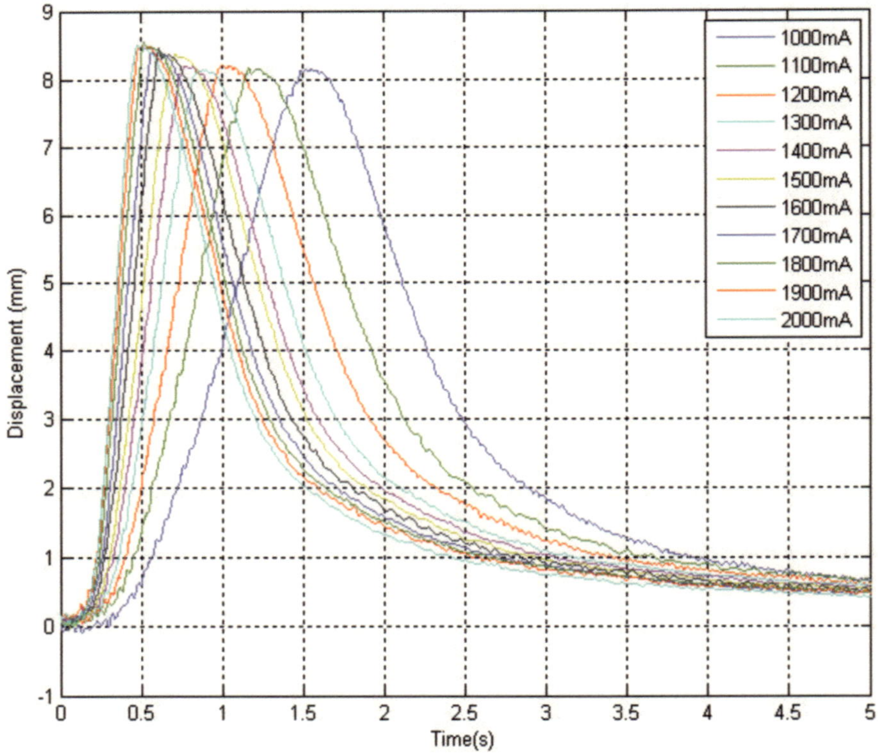

Fig. 14. Motion Profile for different electrical current input signals

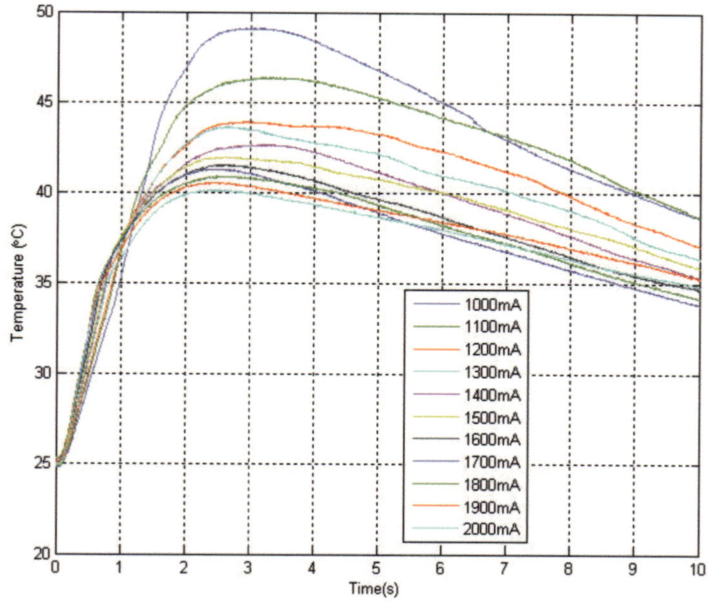

Fig. 15. Temperature Profile for different electrical current input signals (t=10s)

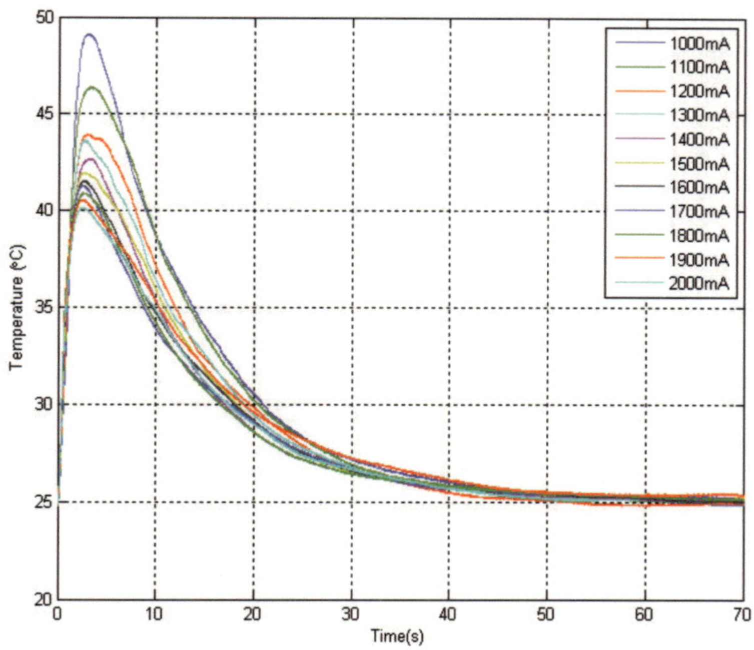

Fig. 16. Temperature Profile for different electrical current input signals (t=70s)

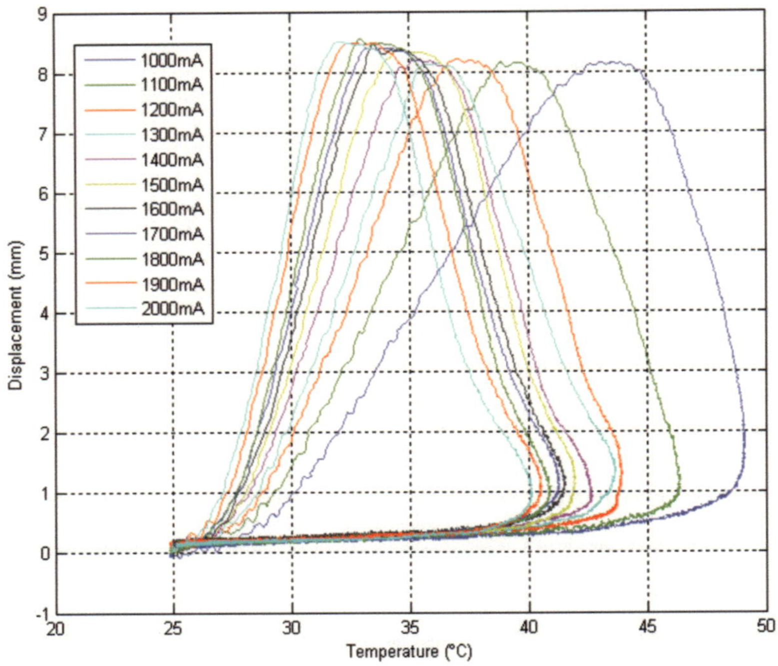

Fig. 17. Displacement vs. Temperature for different electrical current input signals

Next, the performance of the AU is studied for different loads (Fig 18). Supplying the same amperage (Irms = 1400mA), measurements are taken for three different bias loads of the extension spring: Low Load = 5.32N, Average Load = 7N, High Load = 8.68N. From the plots shown below, one may notice a slight faster response for the lighter load. This can be easily explained considering the elasticity of the SMA wire. The elastic strain acts antagonistically to the shape memory strain. So, for a larger load, there is more elastic strain and thus more shape memory strain is needed to cancel it. As a result, there is a longer delay before the load starts to move.

The speed performance of the finger can also be evaluated from this plot. The time needed for extension of the finger is less than 0.9 seconds. Passive return to the rest position is a more slow process: about one second after power is off an approximately 70% of the displacement is restored, while with a 2.5 seconds of cooling the finger flexes about 90% of its maximum range.

As it has been already observed (Teh, 2008), the electrical resistance of the SMA wire varies directly with its length. Indeed, our experiments (Fig. 19) confirm that during its heating phase the resistance drops about 15% with an almost linear relationship with contraction. Thus, measuring the resistance value, position feedback can be provided eliminating the need for additional position sensors.

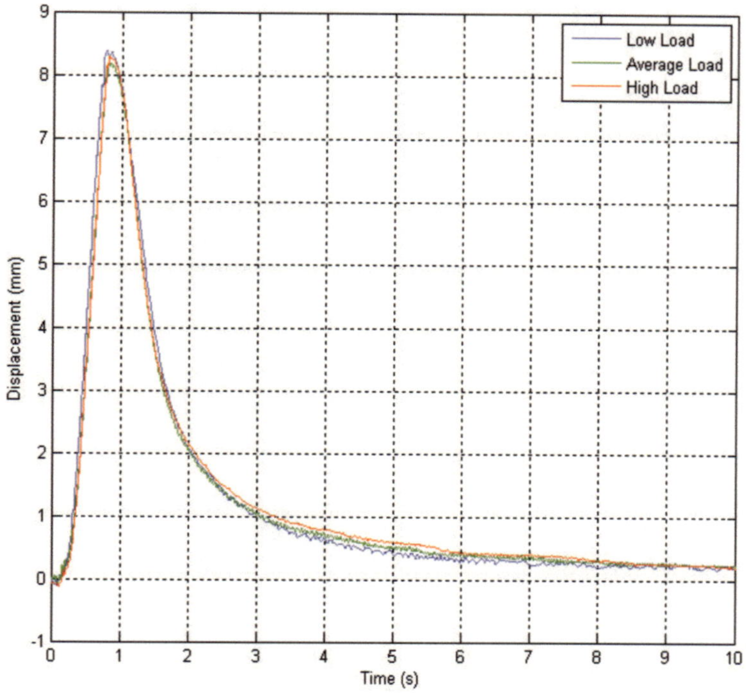

Fig. 18. Motion Profile for different loads

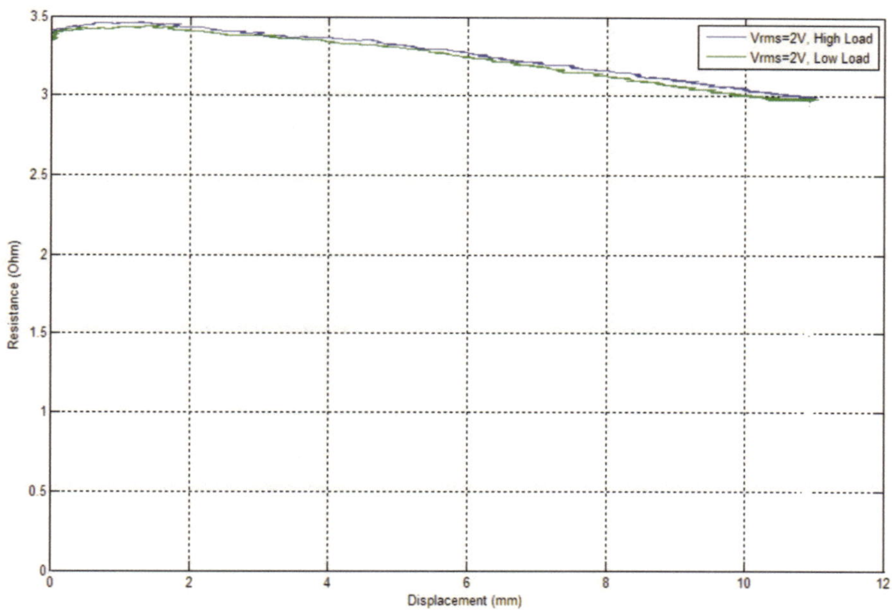

Fig. 19. Resistance vs. Displacement for different loads

Closed-loop Experiments:
In order to assess the performance of a position feedback control to the system, tests are carried out using the control infrastructure shown in Figure 20.

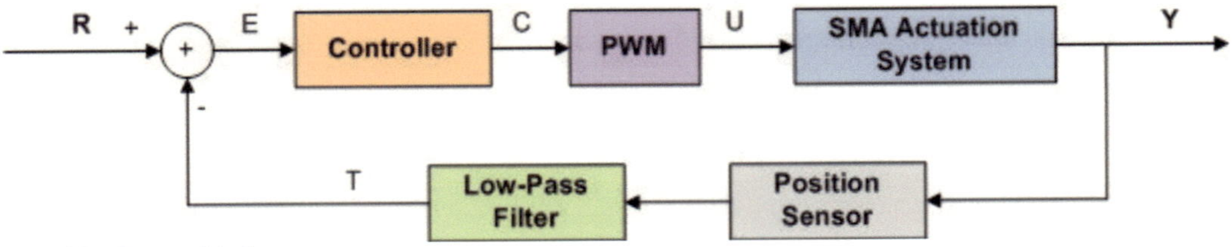

Fig. 20. Control infrastructure

In literature, various control strategies have been developed to cope with the nonlinearities and hysteresis phenomena of the SMA. A thorough review of these strategies can be found in (Rezaeeian et al., 2008; Cocaud et al., 2006; Ashrafiuon et al., 2006). Briefly, researchers have explored linear controllers as well as nonlinear control schemes including fuzzy logic, neural networks, feedback linearization, optimal control and variable structure control. However, the large range of factors affecting SMA behavior (e.g ambient settings, stress, strain and material fatigue) hinder the robustness of the majority of controllers.

As it is underlined in (Cho and Asada, 2006a), due to the bi-stable nature of SMA's phase transition, an on-off type actuation is more suitable than a proportional control. Thus, in our research, an on-off controller is initially tested and its results are presented below. With this type of controller, an electrical current of Irms=1400mA is applied to SMA wire when there is a negative position error (E = R-T) otherwise current is zero. Results for an incremental step input are shown in Figure 21 indicating serious overshoot problems and intense fluctuations around steady-state values. Consequently, it is clear that an ON-OFF controller is incompetent to provide satisfactory responses for our system.

Therefore, other control approaches were examined for our system. After experimentation with various control schemes, a gain-scheduling controller was selected. This is one of the simplest and most intuitive forms of adaptive control. It is an approach to control of non-linear systems using a family of linear controllers, each of which provides satisfactory control for a different operating point of the system. One or more observable variables, called the scheduling variables, are used to determine what operating region the system is currently in and to enable the appropriate linear controller. In our case, position error is the only scheduling variable. The operating regions and gains are tuned empirically.

Experimental results of Figures 22-25 indicate the efficacy of the suggested controller. In Figure 22, finger demonstrates a significant improvement to its performance compared to the ON-OFF controller. Overshoot problems have been eliminated and finger's stability around different set-points has turned to be more robust. In Figures 23-24, the responses for input signal tracking using sinusoidal command signals of 0.15 and 0.05 Hz are shown respectively. In this case, the reference signal is selected regarding the speed performance computed before in open-loop experimentation. Finally, a qualitative study of the robustness of this controller is provided in Figure 25 where several external disturbances are imposed to the system.

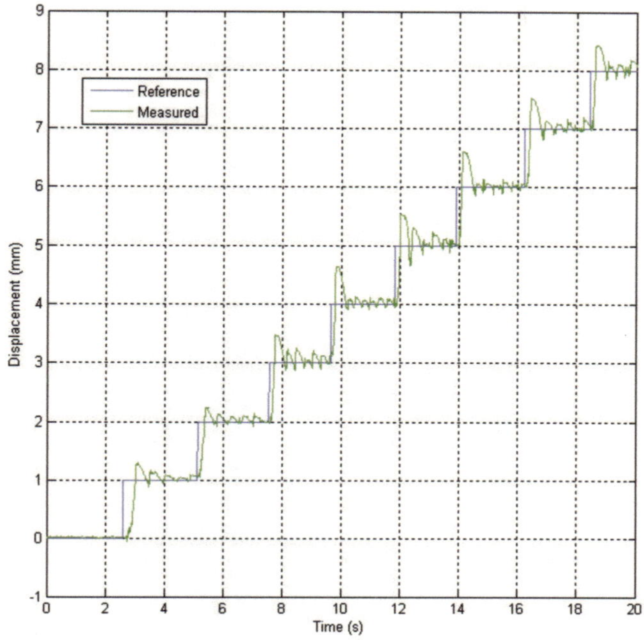

Fig. 21. Position Response to an incremental step input with an ON-OFF controller

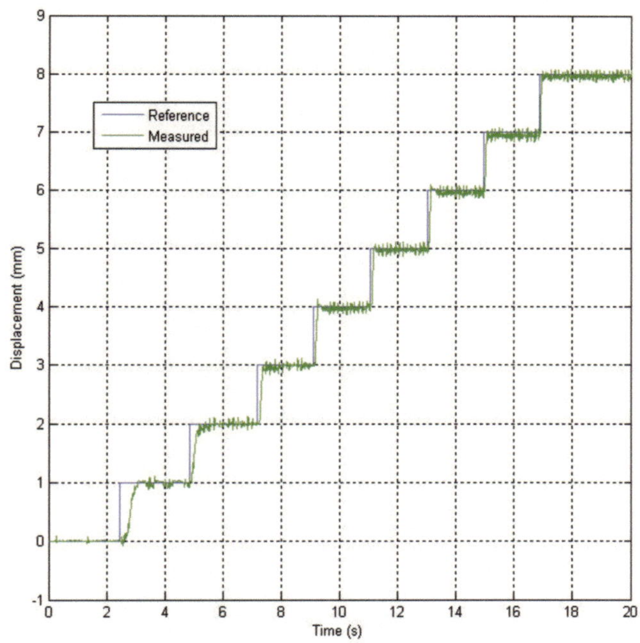

Fig. 22. Position Response to an incremental step input using a gain-scheduled controller

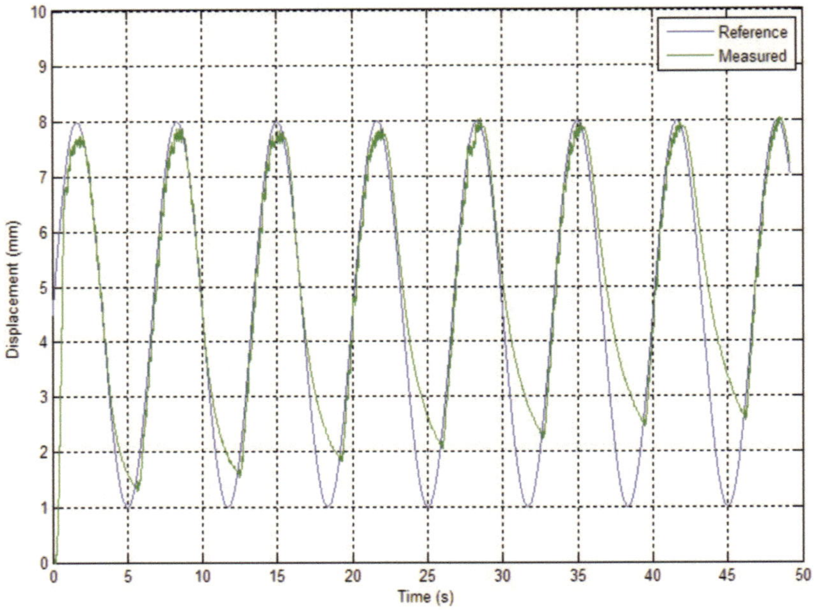

Fig. 23. Position Response to a 0.15Hz sinusoidal input using a gain-scheduled controller

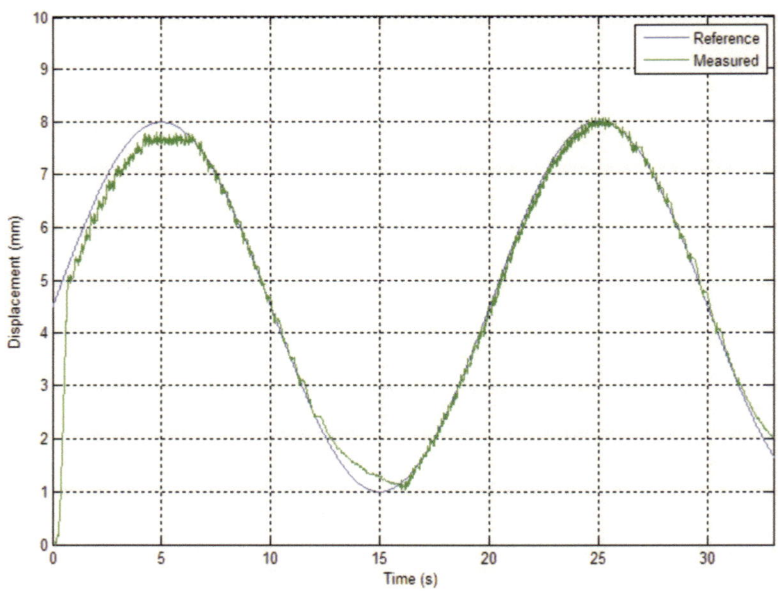

Fig. 24. Position Response to a 0.05Hz sinusoidal input using a gain-scheduled controller

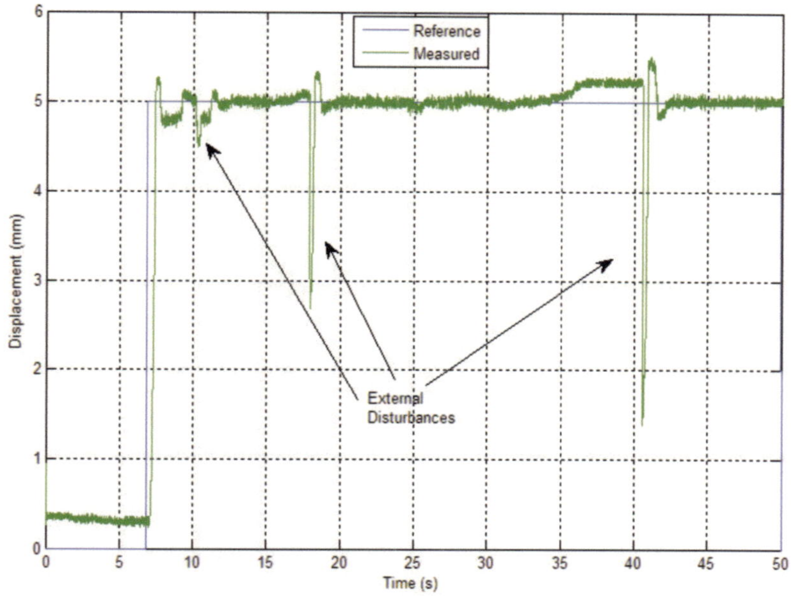

Fig. 25. Finger's Position Response to a step input with external disturbances using a gain-scheduled controller

5. Conclusions & Future Work

It is a common belief that improvement of actuation systems will have a crucial impact on the development of more anthropomorphic and functional prosthetic hands. One of the emerging technologies in the actuation field concerns the utilization of novel, smart material based SMA wires. Their incredibly high power to weight ratio along with their ability to operate with a biologically similar motion suggest that they could be an ideal solution to the size and weight constraints of prosthetics.

In this research, an SMA actuation system integrated in a prosthetic hand was presented. Owing to the lightweight and silent nature of the SMA, this hand can enhance acceptance by the amputees. Performance of one finger was assessed by open and closed loop experiments. The obtained results encourage future work with the hand prototype. Taking advantage of a voluntary opening mechanism in combination with an underactuation approach, the power needed to perform and maintain a grasp is reduced allowing a rechargeable battery source to operate the prosthesis. Moreover, the response time proved to be satisfactory for manipulative tasks since a finger can open and close with almost the average human speed.

Future work will focus on the prosthetic hand's performance. Electronics will be embedded in structure and different control strategies will be evaluated on the base of resistance feedback. Reliability tests of the proposed prosthesis are also planned to be conducted for both static and dynamic grasps, with its compliance and conformability to warrant high grip functionality.

6. References

Abolfathi, P. (2007). "Development of an Instrumented and Powered Exoskeleton for the Rehabilitation of the Hand," *Mechanical and Mechatronic Engineering*. Sydney: University of Sydney

Albrecht, I; Haber, J. & Seidel, H.P. (2003). "Construction and Animation of Anatomically Based Human Hand Models," In: M. Lin D. Breen, Ed., *Eurographics/SIGGRAPH Symposium on Computer Animation*, pp 98-109

Almstrom, C.; Herberts, P. & Korner, L. (1981). "Experience with Swedish Multifunctional Prosthetic Hands Controlled by Pattern Recognition of Multiple Myoelectric Signals," *International Orthopaedics*, Vol. 5, pp 15-21

Ashrafiuon, H.; Eshraghi, M. & Elahinia, M.H. (2006). "Position Control of a Three-link Shape Memory Alloy Actuated Robot," *Journal of Intelligent Material Systems and Structures*, Vol. 17, No. 5, pp 381-392

Biagiotti, L. (2002). "Advanced Robotic Hands: Design and Control Aspects," *Department of Electronics, Computer Science and Systems*: University of Bologna

Brown, A. (2008a). "Why Hands Matter," *Mechanical Engineering*, Vol. 130, No. 7

Brown, J. (2008b). "Technical Characteristics of Flexinol," Dynalloy, Inc.

Bundhoo, V.; Haslam, E.; Birch, B. & Park, E.J. (2008). "A shape memory alloy-based tendon-driven actuation system for biomimetic artificial fingers, part I design and evaluation," *Robotica*, Vol. 27, No. 1, pp 131-146

Bundhoo, V. (2009). "Design and evaluation of a shape memory alloy-based tendon-driven actuation system for biomimetic artificial fingers," *Department of Mechanical Engineering*: University of Victoria

Carrozza, M.C.; Vecchi, F.; Sebastiani, F.; Cappiello, G.; Roccella, S.; Zecca, M.; Lazzarini, R. & Dario, P. (2003). "Experimental analysis of an innovative prosthetic hand with proprioceptive sensors," *2003 IEEE International Conference on Robotics and Automation*, pp 2230-2235

Cho, K.J. & Asada, H. (2006a). "Architecture Design of a Multi-Axis Cellular Actuator Array Using Segmented Binary Control of Shape Memory Alloy," *IEEE Transactions on Robotics*, Vol. 22, No. 4, pp 831 - 843

Cho, K.J.; Rosemarin, J. & Asada H. (2006b). "Design of Vast DOF Artificial Muscle Actuators with a Cellular Array Structure and Its Application to a Five-Fingered Robotic Hand," *Proceedings of the 2006 IEEE International Conference on Robotics and Automation*. Orlando, Florida - May 2006, pp 2214-2219

Cocaud, C.; Price, A.; Jnifene, A. & Naguib, H. (2006). "Position control of an experimental robotic arm driven by artificial muscles based on shape memory alloys," *International Journal of Mechanics and Materials in Design*, Vol. 3, No. 3, pp 223-236

De Laurentis, K.J. & Mavroidis, C. (2002). "Mechanical design of a shape memory alloy actuated prosthetic hand," *Technology and Health Care*, Vol. 10, No. 1, pp 91-106

Del Cura, V.O; Cunha, F.L.; Aguiar, M.L. & Cliquet, A. (2003). "Study of the Different Types of Actuators and Mechanisms for Upper Limb Prostheses," *Artificial Organs*, Vol. 27, No. 6, pp 507–516

Herr, H.; Whiteley, G.P. & Childress, D. (2003). *Cyborg technology – biomimetic orthotic and prosthetic technology*, Bellingham, WA: SPIE – International Society for Optical Engine

Kelly, B.; Pangilinan, P.; Rodriguez, G.; Mipro, R. & Bodeau, V. (2009). "Upper Limb Prosthetics," *Prosthetics*: eMedicine Physical Medicine and Rehabilitation

Kyberd, P. J., Light, C.; Chappell, P. H.; Nightingale, J.M.; Whatley, D. & Evans, M. (2001). "The design of anthropomorphic prosthetic hands: A study of the Southampton Hand," *Robotica*, Vol. 19, pp 593-600

Lake, C. & Dodson, R. (2006). "Progressive Upper Limb Prosthetics," *Physical medicine and rehabilitation clinics of North America*, Vol. 17, No. 1, pp 49-72

Lan, C.C.; Wang, J.H & Fan, C.H. (2009). "Optimal design of rotary manipulators using shape memory alloy wire actuated flexures," *Sensors and Actuators A: Physical*, Vol. 153, pp 258-266

Maeno, T. & Hino, T. (2006). "Miniature five-fingered robot hand driven by shape memory alloy actuators," *12th IASTED International Conference*. Honolulu, Hawaii, USA pp 174-179

Mavroidis, C.; Pfeiffer, C. & Mosley, M.J. (2000). "Conventional Actuators, Shape Memory Alloys, and Electrorheological Fluids," In: Y. Bar-Cohen, Ed., *Automation, Miniature Robotics & Sensors for Non-Destructive Testing & Evaluation*: The American Society for Nondestructive Testing, Inc. (ASNT), pp 189-214

O'Toole, K. T. & McGrath, M.M.(2007) "Mechanical Design and Theoretical Analysis of a Four Fingered Prosthetic Hand Incorporating Embedded SMA Bundle Actuators," *Engineering and Technology*, Vol. 25, pp 142-149

Price, A.; Jnifene, A. & Naguib, H. (2007). "Design and control of a shape memory alloy based dexterous robot hand," *Smart Materials and Structures*, Vol. 16, No. 4, pp 1401-1414

Rezaeeian, A.; Aghil, Y.; Shasti, B. & Doosthoseini A. (2008). "ANFIS Modeling and Feedforward Control of Shape Memory Alloy Actuators," *International Journal of Mathematical Models and Methods in Applied Sciences*, Vol. 2, No. 2, pp 228-235

Schulz, S., Pylatiuk, C.; Kargov, A.; Oberle, R. & Bretthauer, G. (2004). "Progress in the development of anthropomorphic fluidic hands for a humanoid robot," *4th International Conference on Humanoid Robots*. Santa Monica, California, USA, pp 566-575

Teh, Y.H. (2008). "Fast, Accurate Force and Position Control of Shape Memory Alloy Actuators," *Department of Information Engineering*

Trost, F.J. & Rowe, D. (1992). "Upper-Limb Deficiencies: Externally Powered Prostheses," In: Michael JW Bowker HK, Ed., *Atlas of Limb Prosthetics: Surgical, Prosthetic, and Rehabilitation Principles*. Rosemont, IL: American Academy of Orthopedic Surgeons.

Yang, K. & Wang Y. (2008). "Design, drive and control of a novel SMA-actuated humanoid flexible gripper," *Journal of Mechanical Science and Technology*, Vol. 22, pp 895-904

6

Active Bending Catheter and Endoscope Using Shape Memory Alloy Actuators

Yoichi Haga[1], Takashi Mineta[2], Wataru Makishi[1],
Tadao Matsunaga[1] and Masayoshi Esashi[1]
[1]Tohoku University
[2]Yamagata University
Japan

1. Introduction

Shape memory alloys (SMAs) are excellent materials for microactuators which can generate large force and large displacement. Shape memory alloy can recover its original shape after it is deformed over its elastic limit by external force. Several mechanisms with SMA micro-coil actuators, for example, bending, extension/contraction, torsional and stiffness control mechanisms have been developed. The principles underlying these mechanisms and the structure of each mechanism are detailed and medical applications, namely, catheters and endoscopes are presented.

2. SMA (Shape Memory Alloy)

2.1 Principle

Shape memory alloy (SMA) recovers previously memorized shape upon heating. The alloy has to be deformed in its Martensitic low temperature phase and subsequently heated to the Austenitc high temperature phase. The SMA generates a large deformation and large force during the phase transformation.

The response of SMA actuators tends to be slow due to the heat capacity of the actuator, resulting in lengthy heating and cooling. Because deformation of SMA is caused by a crystalline phase transformation which occurs without any diffusion of substances (Otsuka & Wayman, 1998), an SMA actuator can deform rapidly when it is miniaturized thus reducing its heat capacity.

The material of most bulk SMA is TiNi alloy. Because of the poor mechanical workability of TiNi alloy which results in high cost and defects in SMA, choice of shapes of bulk SMA is limited.

2.2 Fabrication and assembly of SMA microactuator

As a micromachining method for TiNi SMA, laser cutting is practical because it is fast and allows more complex patterns (Reynaerts et al., 1999). To prevent deterioration due to heat generation by the laser, appropriate laser power and cutting speed should be selected. A femtosecond laser is effective to prevent heat generation, but the cost of the system is relatively high.

Electrochemical etching is one of the most suitable methods to etch TiNi SMA with a high etch rate. Compared to conventional chemical etching, electrochemical etching has features including high etching rate, low side etching and a smooth etched surface. An etch factor (etched depth/under-cut) higher than 1.5 can be obtained. Another advantage of electrochemical etching is that normal photoresist can be easily used as an etching mask. Electrochemical etching can be carried out using a simple setup, as shown in Fig. 1 (Mineta et al., 2000; Mineta et al. 2002).

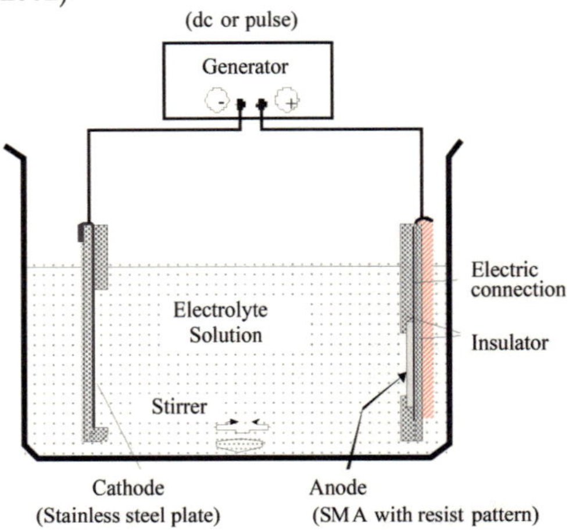

(a) Setup for electrochemical etching of SMA

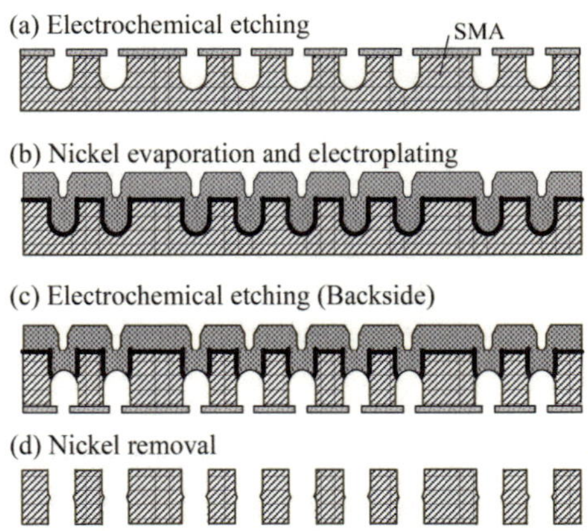

(b) Double-side electrochemical etching of TiNi sheet using a nickel dummy layer

Fig. 1. Electrochemical etching of SMA sheet

Shape memory alloy coated with a photoresist is used as an anode. The anode and a counter cathode, for example a stainless-steel plate, were facing each other with a gap of several tens of millimeters in the electrolyte solution. A solution of 5 vol % (about 1 mol/L) H_2SO_4-methanol has been used as an electrolyte for electrochemical etching of SMA (Allen & Chen, 1997) (Kohl et al., 1994). When a direct-current (DC) voltage, typically 8-10 V, is applied between the electrodes, the SMA anode is etched electrochemically. In addition to continuous voltage, pulse voltage is also useful for the electrochemical etching.

Through-layer etching is important for the precise fabrication of micro actuators from an SMA sheet. During over-etching, the remaining SMA patterns tend to be etched non-uniformly due to imbalance of the electrolytic current distribution. The etching proceeds at the places where the current is concentrated, while it stops at other places. In order to overcome the problem of non-uniform sheet-through etching, a conductive dummy layer of Ni or Cu, which is formed on the back side of the SMA sheet previously, is very effective in maintaining a uniform current distribution during the over-etching. The dummy layers of Ni and Cu can be easily removed in concentrated nitric acid without damage to the SMA surface (Mineta et al., 2000). Meandering SMA micro actuators fabricated by electrochemical etching of 40 μm thick SMA sheet is shown in Fig. 2 (a).

(a) Electrtochemical etching of SMA sheet (40 μm thick, Double-side etching)

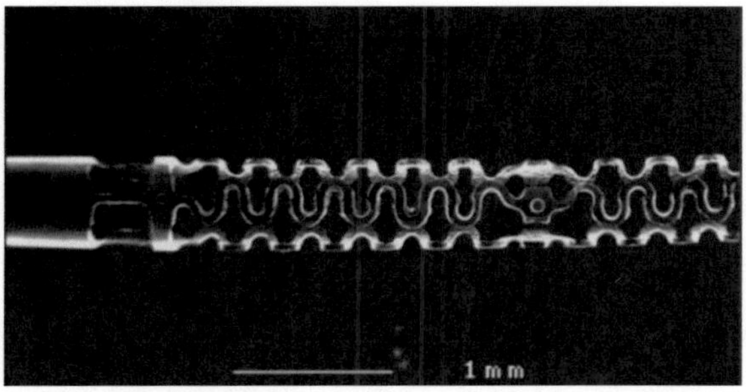

(b) Tubular SMA actuator-unit electrochemically etched from a SMA tube (outer/inner dia.: 600/500 μm)

Fig. 2. Meandering SMA micro actuators fabricated using electrochemical etching

Electrochemical etching also can be used for fabrication of non-planar SMA structures. Fig. 2 (b) shows a tubular SMA actuator unit for an active catheter bending mechanism, in which meandering-shaped SMA actuators are formed. An SMA tube, covered with a photoresist pattern formed by non-planar photolithography, is electrochemically etched in a cylindrical vessel with a cylindrical cathode, resulting in the fabrication of a tubular SMA actuator-unit (Mineta et al., 2004).

2.3 Assembly of SMA microactuator

For purposes of mechanical fixation of SMA or parts with SMA, several assembly methods, for example, mechanical fixation, adhesion, welding and soldering are used.

The disadvantages of mechanical fixation are the requirement of large connection areas, impreciseness of length adjustment, low productivity because of the handmade process and fracture susceptibility in the fixation area because of local deformation and local stress of SMA.

A disadvantage of adhesion is that considerable time is necessary to make all connections when these connections have to be formed individually. Cyanoacrylates can be cured in a shorter time than epoxy resin but are relatively brittle. Adhesive tape (single sided or double sided) can be used for temporary fixation.

Regarding the heating process for heat-curable adhesive, soldering or welding, the temperature limit and heating time in the assembly should be considered to prevent deterioration of the shape memory effect. As SMA deformed from its memorized shape tends to be restored to its memorized shape during the heating process, a certain fixation method of the SMA is required to prevent the restoration.

Using polymer material for intermediate adhesion, not only a nonconductive adhesion part but also a conductive adhesion part using conductive resin is employed. As the nonconductive adhesive part, epoxy resin (Mineta et al., 2000) and UV curable resin (Ballandras et al., 1997) have been utilized. As the conductive adhesive part, epoxy resin containing silver powder (Mineta et al., 2000) or UV curable conductive epoxy resin have been utilized (Lim et al, 1996). Laser assisted deposition has been carried out for fixing an SMA coil to silicon parts for active bending catheters (Lim et al, 1996), vaporized polymer source material being selectively polymerized and deposited on the UV light-irradiated area. As considerable time is necessary to make all connections when these connections have to be formed individually, batch assembly method using nickel electroplating and acrylic resin electrodeposition has been developed (Haga et al., 2000). Fabrication is carried out as follows. The SMA coil and metal spring are coated with polymer and the polymer is partially removed by laser ablation for partial exposure of the SMA surface and metal surface. UV curable acrylic resin is locally deposited by electroplating between the exposed SMA surface and the exposed metal surface as shown in Fig. 3. Oxide film on the surface of TiNi alloy should be removed to achieve uniformity of electroplating.

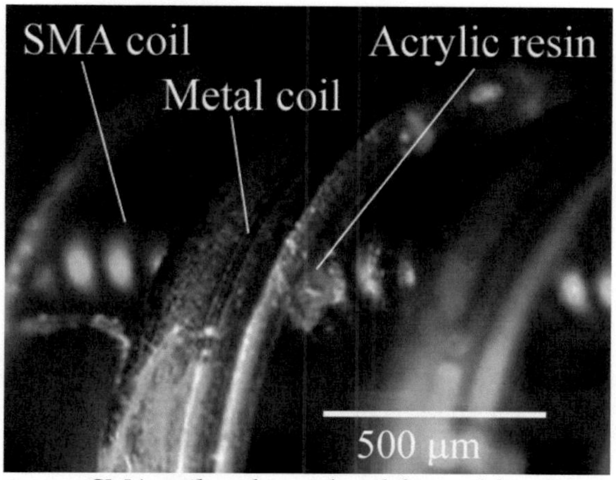

Fig. 3. Connection between SMA coil and metal coil formed by electrodeposition of acrylic resin

2.4 Actuation of SMA microactuator

Large scale bulk SMA actuators tend to have a slow response due to the large heat capacity of the actuator itself, which results in lengthy heating and cooling cycles. An SMA actuator can be rapidly deformed when it is miniaturized so as to reduce its heat capacity. An SMA micro actuator can be heated above its phase transformation temperature rapidly by small thermal energy and can be rapidly cooled down by ambient air without the addition of a cooling mechanism.

2.4.1 Direct heating

TiNi SMA can be actuated by Joule heat created by direct flow of an electrical current into the SMA when electrical resistance of the SMA is relatively high. The advantage of the actuation method is its simple structure without any heater or cooler. Disadvantages of direct heating are the requirement of high electrical resistance and electrical connections with low electrical resistance between the SMA and dissimilar conductive material.

Relatively high electrical resistance of SMA is required to avoid a high electrical current for actuating the SMA. Though the small or thin structure of SMA has high electrical resistance, it cannot generate high power. An array of small or thin SMA structures connected as a series circuit can solve this problem, but the structure is relatively complex.

Electrical connections with low electrical resistance between SMA and dissimilar conductive material, for example, copper as electrical lead wires, is required to avoid undesirable consumption of electrical current for heat generation at the connection.

2.4.2 Indirect heating

Indirect heating of an SMA actuator by additional heater lines with high resistance is advantageous for reducing the electric power needed for actuation. To realize indirect heating, it is necessary to select an appropriate mechanical design and fabrication processes of the heater and insulator so that they can deform without fracture during the SMA actuation. For example, a thin Ni film (0.6 μm) on a Parylen-coated SMA coil is used as an indirect heater (Lim et al, 1996). The Parylene layer (1 μm thick) is an insulator between the Ni layer and the SMA surface. Parylene can be deposited uniformly in a vacuum on the entire surface of the SMA coil. A thin Ni heater layer is formed by electroless plating on the Parylene. No damage is observed in the Ni or Parylene films after 3000 cycles of prolonged experiments with applied power of 80 mW at 2 % strain of SMA coil.

3. Active Catheter

3.1 Catheter and motion control mechanism

To inject contrast medium or medicine into the blood vessel or to measure blood pressure locally, a catheter, which is a hollow flexible small tube, is used. Catheter-based procedures (catheterization) enable doctors to access almost every diseased site which needs to be checked and treated via a blood vessel. Doctors control the tip of the catheter by moving its proximal part from outside of the body. Depending on blood vessel size, the catheters are approximately 0.3-3.0 mm in diameter and 1.5 m in length. As doctors must control the tip of the catheter from outside the body, operations with catheters require considerable skill, particularly when the blood vessel has a loop or complex configuration. Thus, there is a demand for an active catheter which moves like a snake in the blood vessel. Precise and safe manipulation of the catheter is realized if doctors can control the motion of the catheter from outside the body. As back-and-forth operation of the catheter from outside the body is relatively easy, a self-propelling mechanism is not necessarily demanded. Consequently, almost all controllable steering mechanisms are installed near the tip of the catheter. Active catheters which have micro-actuators at the tip have been developed for use in every part of the body. The tip can be controlled from outside the body and moves like a snake if many micro-actuators are distributed and properly controlled. Not only bending motion, but also torsional and precise back-and-forth motion is useful for manipulation of the tip of the catheter.

3.2 Active bending mechanism using SMA microactuator

An active catheter which has a bending mechanism realized by thin TiNi SMA plates has been developed (McCoy, 1985) (Kaneko, 1995). One of the SMA plates fixed along the side of the catheter bends when the plate is heated by application of a current to restore its memorized shape. To realize precise control of bending motion by feedback control using temperature monitoring, a component which has heaters, temperature sensors and circuits on a flexible polyimide film substrate has been fabricated using MEMS technology (S. Kaneko, 1995). The film components are fixed on the TiNi SMA plates placed on the sides of the catheter. To realize multi-directional bending, pairs of film components and SMA plates are fixed serially on alternating sides of the catheter. The external diameter of such a fabricated multi-joint catheter is 1 mm. The SMA plates utilized for this catheter are two directionally memorized SMAs, the shape of which is restored by cooling. A bending mechanism with a diameter of 15 mm has been developed for gastrointestinal application (Reynaerts et al., 1999). This mechanism consists of a stack of elements with an SMA plate at the center of each element. When the SMA plates are actuated by the application of current, the whole structure bends in two directions like a vertebra of the human body.

Shape memory alloy wires are also utilized for active catheters (Fukuda et. al, 1994) (Takizawa et al., 1999) (Lim et al. 1999). When the SMA wires are embedded in the catheter eccentrically and heated above a certain transformation temperature by direct application of an electrical current to the SMA wire, the wire actuator bends the catheter by contraction of its length. The contraction length of the SMA wire, however, is relatively short and SMA wires fixed to the opposite side tend to restrict bending motion since the wires have to bend and stretch passively. Consequently, it is difficult to realize bending with a small radius of curvature.

Shape memory alloy micro-coil actuators are used to obtain a large bending motion and to realize multi-joint actuation as shown in Fig. 4. Shape memory alloy coils enable multi-directional bending with a large bending angle because the SMA coil actuators fixed at the opposite side can be passively extended. Shape memory alloy coil actuators can also be actuated by direct application of current when the wire diameter of the coil is small. Three SMA coil actuators, which are extended (from their memorized shapes) and fixed in the catheter, contract and bend in several directions. Multi-joint and multi-directional active catheters using silicon-glass link structures have been fabricated. Three SMA coil actuators are fixed between two silicon-glass link structures at intervals of 120 degrees. Many joints are serially connected and each joint can bend in any direction. The external diameter of the fabricated catheter is 2.7 mm (Lim et. al, 1996).

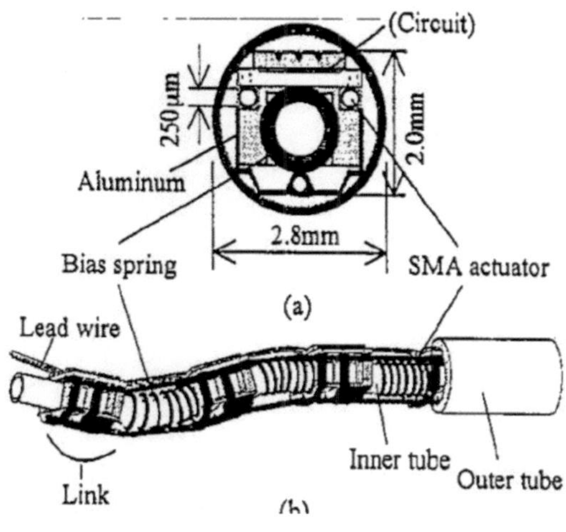

Fig. 4. Active bending catheter using SMA coils

By locating SMA coil actuators inside the liner coil and fixing them to the liner coil directly, link structures can be eliminated and the liner coil can be used as a skeleton and a bias spring (Fig. 5). Furthermore, every part works as a joint and hence bends more flexibly than that with links (Haga et. al, 1998). As the catheters for use in blood vessels are disposable to avoid blood infection, active catheters need to be fabricated at low cost. To solve this problem, batch assembly methods have been developed. One is silicon wafer-level batch fabrication with a silicon joint structure (Mineta et. al, 1999) and the other is fabrication using nickel electroplating and acrylic resin electrodeposition (Haga et. al, 2000). Fabrication using electroplating and deposition is carried out as follows. The SMA coil and metal spring are coated with polymer and the polymer is partially removed by laser ablation for partial exposure of the SMA surface and metal surface. UV curable acrylic resin is locally deposited by electroplating between the exposed SMA surface and the exposed metal surface. This novel method enables low cost batch assembly and a small external diameter.

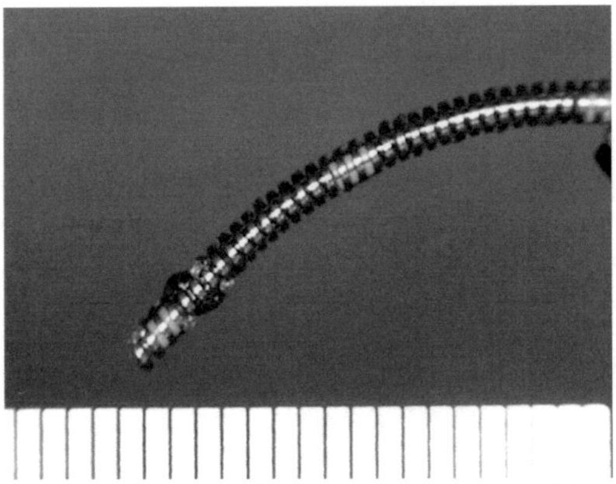

Fig. 5. Active bending catheter fabricated using electrodeposition (external diameter, 1.6 mm)

Instead of a coil shape, flat meandering-shaped SMA actuators are also used for active catheters. The flat shape meets both demands of small external diameter and large working channel by realizing a thin wall (Lim & Lee, 1999). A batch fabrication method for flat meandering SMA actuators from an SMA (TiNi) sheet has been developed using electrochemical pulsed etching for active catheters (Mineta et. al, 2000). Furthermore, to simplify the assembly process of the SMA actuators, three meandering actuators are formed in an SMA pipe using electrochemical etching mentioned above.

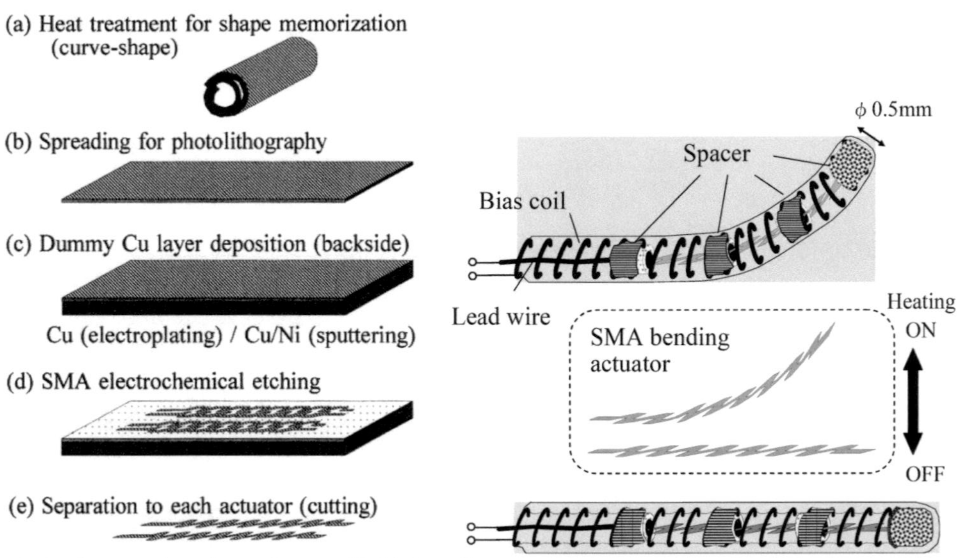

Fig. 6. Active guide wire using an SMA bending actuator

Guide wires are wires which have a relatively flexible tip and are used for guidance of the catheter. An active bending guide wire has been fabricated using a flat meandering actuator (Fig. 6). The guide wire is bent by an electrical current supplied to the actuator because the actuator has a memorized curved shape (Mineta et. al, 2002).

3.3 Other motion mechanisms

By changing the deformation shape of the SMA and the configuration of the mechanism, several motions can be realized, not only bending motion, but also torsional and extension motions (Fig. 7).

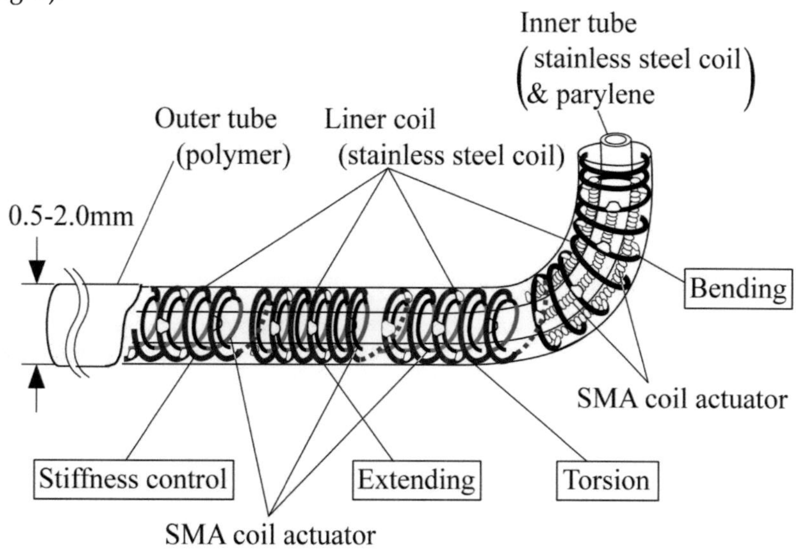

Fig. 7. Bending, torsional and extending active catheter using SMA coil actuators

3.3.1 Torsional mechanism

A torsional mechanism has been developed for medical application. For conventional manipulation of catheters through branches of blood vessels, a J-shaped tip of the catheter or guide wire, which is used for guidance of the catheter, is torsionally rotated from outside the body. However, the torque cannot be transmitted with good control when a blood vessel has a loop or complex configuration. This problem can be solved by installing an active torsional mechanism at the tip of the catheter or the guide wire. The structure of the torsional mechanism using the SMA micro-coil actuator consists of a metal spring coil and a twisted SMA coil fixed coaxially inside the metal coil. The metal coil plays the roles of a bias spring and a lead wire. When the SMA coil is heated above a certain transition temperature by an electrical current, the SMA coil is untwisted. Conversely, the metal coil twists (turns back) the catheter when the electrical current is turned off. The external diameter of the fabricated torsional mechanism is 1.3 mm without an outer tube and its length is 7 mm (Fig. 8). A torsional rotation of 70 degrees can be obtained with a current of 80 mA (Haga et al., 2000). Fabrication is carried out using electrodeposition of acrylic resin.

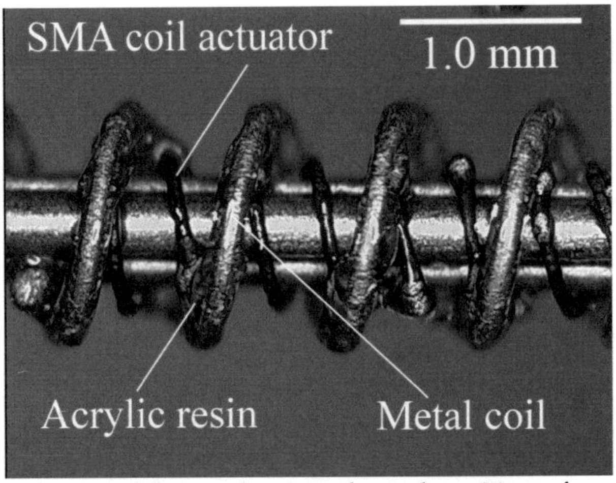

Fig. 8. Torsional mechanism fabricated using electrodeposition of acrylic resin (external diameter, 1.3 mm)

3.3.2 Extension/contraction mechanism

An extension/contraction mechanism has been developed for back-and-forth motion for precise manipulation of catheters, endoscopes and guide wires. This mechanism has also been employed for a dynamic pin display for the visually impaired or sightless.

There are mainly two ways for realizing an extension/contraction mechanism with SMA micro-coil actuators. When a compressed SMA micro-coil actuator is heated by an electrical current, the SMA extends, causing an extending motion. Conversely, an extended SMA micro-coil actuator contracts, causing contracting motion, when the SMA is heated by an electrical current.

3.3.3 Stiffness control mechanism and laser machining of TiNi SMA tube

Application of a stiffness control mechanism is expected to improve manipulation of medical instruments (endoscopes, catheters or guide wires) by changing the stiffness of the shaft of the instrument. A hard shaft is preferable during passage through a relatively safe site, while a soft, flexible shaft is preferable during passage in a relatively dangerous site, for example, a curved area or narrow passage. The stiffness control mechanism is also expected to be useful during the therapeutic operation of intravascular occlusion (recanalization) with guide wires. Chronic total occlusion of a blood vessel is treated by pushing a hard-tipped guide wire through the occlusion. Presently, several shape and stiffness guide wires have to be selected and used for precise and safe intravascular treatment. A guide wire whose shape and pushability can be controlled would be useful for such purposes.

Stiffness can be changed when the shaft of a medical instrument is equipped with an SMA micro-coil actuator in the state of its memorized shape. When the SMA is heated by an electrical current, its stiffness changes due to the shape memory effect.

For this purpose, a spring-shaped SMA micro-coil with a square cross section is preferable. Such an SMA micro-coil has been fabricated by spiral cutting of a TiNi SMA tube with a femtosecond laser. The conventional method of fabricating an SMA micro-coil is coiling of SMA wire. This method however, is not suitable for making a spring-shaped SMA micro-coil with a square cross section. Use of a femtosecond laser enables micromachining of the memorized SMA without deterioration of the shape memory effect because its pulse width is very short (10^{-15} sec) and heat generation is sufficiently reduced during laser machining. The external diameter of the fabricated TiNi SMA micro-coil actuator is 0.9 mm and the size of its wire is 0.07 x 0.25 mm. The pitch of the turns of the micro-coil is 0.4 mm. Stiffness change of the 10-mm-long SMA coil was evaluated by pushing it in the longitudinal direction with a force gauge until flexion occurred. A change of 20 mN was measured by a flowing electrical current of 150 mA.

3.4 Communication and control circuit

A major problem with active catheters which has many joints and functions is the necessity of too many lead wires to control each SMA actuator. To minimize the number of lead wires, flexible polyimide-based integrated complementary metal oxide semiconductor (CMOS) interface circuits for communication and control (C&C) have been developed and utilized in active catheters. To reduce the system size and simplify the assembly work, the C&C integrated circuit (IC) and three lead wires are fabricated on the same substrate using a CMOS-compatible polyimide-based process (Fig. 9). The outer diameter of the fabricated active catheter without an outer tube is approximately 2.0 mm (Park et al., 1996).

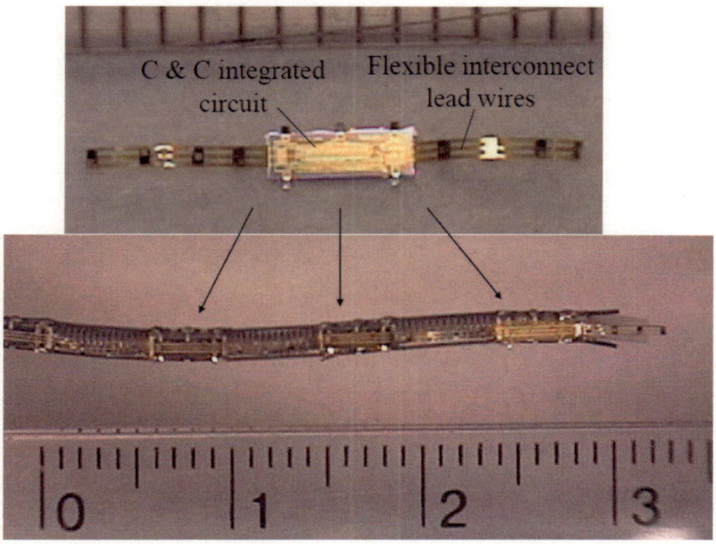

Fig. 9. Torsional mechanism fabricated using electrodeposition of acrylic resin (external diameter, 1.3 mm)

4. Active Bending Intestinal Tube

Intestinal obstruction (ileus) is a serious passage disorder in the intestines which causes acute pain. There are two methods for treatment of ileus. One is abdominal surgery and the other relatively minor method is nonoperative treatment by insertion of a long intestinal tube (LIT) made of silicone rubber to the intestine and depressurization by continuous suction from outside the body. Half of the patients are cured with the latter treatment in Japan. As shown, a doctor inserts the LIT into the patient's nasal cavity and pushes it forward from outside the body, while monitoring its position using X-ray fluoroscopy. In insertion of the tube, it is difficult to pass the lower opening of the stomach (pylorus) because of its narrowness. Fig. 10 shows the pylorus and the tip of the LIT. The external diameter of the tip is 6 mm. Doctors must be skilful and sufficiently experienced to successfully manipulate and insert the tube into the intestine (Hachisuka et al., 1991).

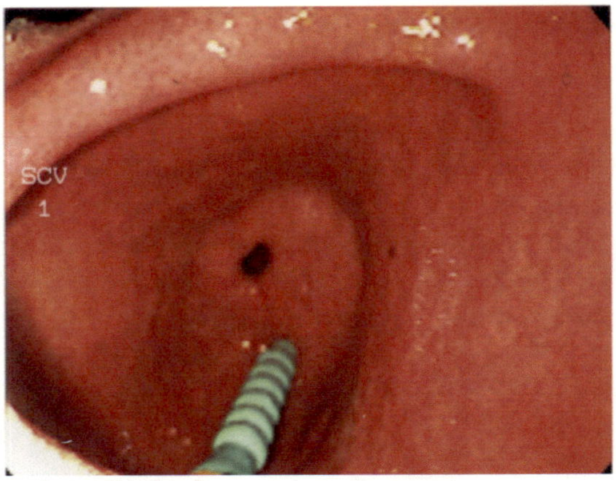

Fig. 10. Pylorus and long intestinal tube

An LIT that has stainless steel weights at the tip is most widely used. This tube utilizes gravity to facilitate manipulation of the tip. However, a patient must endure pain when he or she changes position. Furthermore, accuracy of positioning of the tip is insufficient. In the endoscopic procedure, doctors can control the motion of the tip of the endoscope by manipulating wires in the shaft of the endoscope from outside the body. However, this method cannot be applied to control the tip of the LIT because of its length (over 3 m) and buckling of the silicone rubber shaft due to its softness. For precise manipulation of the tip of the LIT, an active bending LIT incorporating an SMA micro-coil actuator has been developed for easy passage of the pylorus (Mizushima et al., 2004).

Fig. 11 shows the structure and principle of this LIT. Although it has a one-directional bending mechanism, doctors can control the bending direction by rotating the shaft from outside the body. The active bending element consists of a silicone inner tube, polymer links, a TiNi SMA micro-coil actuator placed along the inner tube and a silicone outer tube. The external diameter of the employed TiNi SMA micro-coil actuator is 0.3 mm and the diameter of its wire is 0.07 mm. The SMA actuator is electrically connected to the controller by two electrical lead wires in the small lumen of the shaft.

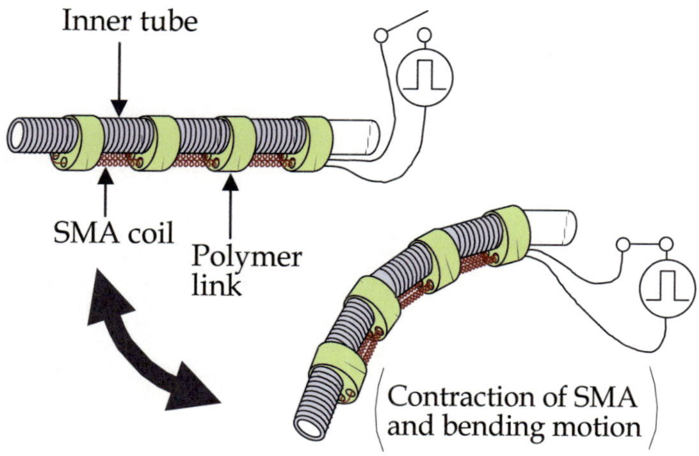

Fig. 11. Structure of active bending element of long intestinal tube

Fig. 12 shows the tip of the fabricated active bending LIT. The active bending LIT system consists of a one-directional 40-mm-long active bending tip with an external diameter of 6 mm, a 3-m-long main silicone tube with an external diameter of 5 mm and a battery powered controller for controlling the bending motion of the tip. The active bending tip is surrounded by a silicone rubber tube with six ring weights made of stainless steel for the conventional manipulation function using gravity, in addition to the active bending mechanism. Its bending angle is controlled by changing the duty ratio by pulse width modulation (PWM). Doctors can easily control the bending motion of the tip of the tube with a stick-type controller. The tube has a through hole as a working channel for insertion of a guide wire or injection of fluids (medicine or contrast medium). The active bending tip bends within 1 sec in 25°C air and maintains its bent shape by the flow of an electrical driving current. The maximum bending angle of the tip is 110 degrees and the radius of curvature is about 20 mm. The bending tip is flexible even if an electrical driving current is supplied and bends because of softness of the SMA micro-coil. A soft tip is preferable for safety operation in the human body. Stiffness of the tip can be evaluated by pushing in the longitudinal direction with a force gauge until flexion occurs. The measured force of a commercialized LIT was found to be 55.2 gf and that of the fabricated active bending LIT was 23.7 gf.

It is desirable that the surface temperature of any medical instrument used in the human body be 41°C or less. The surface temperature of the tip during active bending motion was measured under conditions of 38°C and 95% humidity simulating those in the stomach and the intestine. It takes 45 seconds for the surface temperature to rise over 41°C when the bending angle is maintained at 30 degrees.

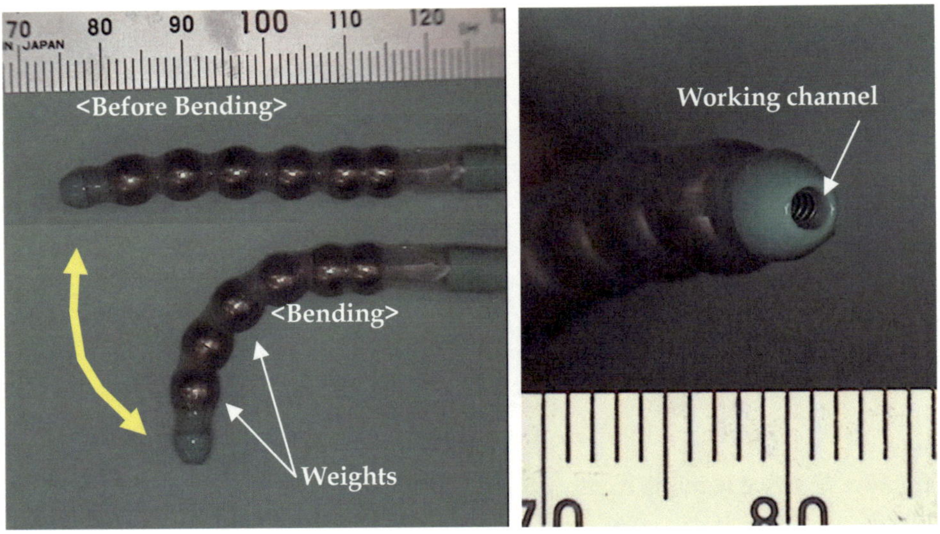

Fig. 12. Active bending tip of long intestinal tube (LIT)

5. Active Bending Electric Endoscope

Several flexible endoscopes are used in the esophagus, stomach, duodenum (gastrointestinal endoscopy), colon (colonoscopy), and ureter and kidney (ureteropyeloscopy). These endoscopes can be bent (deflected) by manipulation of wires from outside the human body. The tip of a conventional endoscope has a bending mechanism using wire traction from outside the body of a patient. The shaft of an endoscope should be relatively hard to avoid buckling by wire traction. Therefore, precise operation of the endoscope is difficult in complex shaped areas such as the intestine which has a complex structure and makes it difficult to manipulate the endoscope accurately.

One of the purposes of active bending endoscopes is insertion deep into the digestive tract, for example small intestine. The other purpose of active bending endoscopes is inspection deep area of the human body where conventional endoscopes which have relatively hard shaft is difficult to achieve.

A capsule endoscope can achieve inspection deep area of the human body but the capsule endoscope cannot be precisely controlled in position, especially in wide areas and the double-balloon endoscopes cannot be thin enough and be inserted from nasal cavity.

For colonoscopy, active steerable endoscopes which diameter is relatively big have been developed. To pass through the colon, which meanders with a small radius, a steerable endoscope with a diameter of 13 mm and incorporating SMA coil actuators has been developed. The SMA coil actuator contracts when heated by the application of an electrical current from outside the body. Consequently, contraction of the SMA coil bends the endoscope. The external diameter of the utilized SMA coil actuator is 1.0 mm and the diameter of its wire is 0.2 mm. Five bending segments are serially connected and controlled synchronously with an external servomotor which moves the endoscope back and forth from outside the body (Ikuta et al., 1998). Unfortunately the bending mechanism doesn't have any imager. A steerable tip with a CMOS (Complementary Metal Oxide Semiconductor) imager has also been developed using SMA coil actuators. The external diameter of the utilized SMA coil actuator is 0.7 mm and the diameter of its wire is 0.22 mm (Menciassi et al., 2002).

We have developed an active bending electric endoscope using SMA coil microactuators with a CCD (Charge Coupled Device) imager. The external diameter of the fabricated endoscope is 5.5 mm, the external diameter of the utilized SMA coil actuator is 0.3 mm and the diameter of its wire is 0.075 mm (Makishi et al., 2000). Three SMA coil actuators, which are extended (from their memorized shapes) and fixed in the endoscope, contract and bend in several directions. Bending angle and direction is controlled using joystick controller. The fabricated bending mechanism makes the shaft of the endoscope soft and thin (external diameter 5.5 mm), because SMA coil actuators are soft and traction wire is not used.

For more thin and short rigid area of the active bending endoscopea FPC (Flexible Circuit Board) like spiral cut cylindrical substrate for thin endoscope has been fabricated using laser processing on cylindrical substrates (Fig. 13). The fabricated cylindrical substrate consists of a spiral flexible part as a lead wire and a polymer spring of bending mechanism and circuit pattern for mounting CCD imager related parts. Fabrication process is shown in Fig. 14 (a). Metallization and patterning have been performed on a polyimide tube with 2 mm external diameter using maskless lithography techniques with a laser exposure/ablation system as shown in Fig. 14 (b). Cu layer is formed as a seed layer and a thick copper layer is electroplated after exposure and development of the resist. Fig. 15 shows a mounted CCD imager and electric components at a tip of the substrate. White LEDs for light guide are mounted at the tip of endoscope. As shown in Fig. 15 (b) the fabricated endoscope has a working channel at the side wall. The external diameter and length of the CCD imager part are 3.9 mm and 15 mm respectively. An inner diameter of working channel which is located on a side is 1.5mm. Bending motion and inspection using mounted CCD is demonstrated in a colon model as shown in Fig. 16. The maximum bending angle of the fabricated endoscope is 98 degrees (Curvature radius is 19.3 mm at 120 mA).

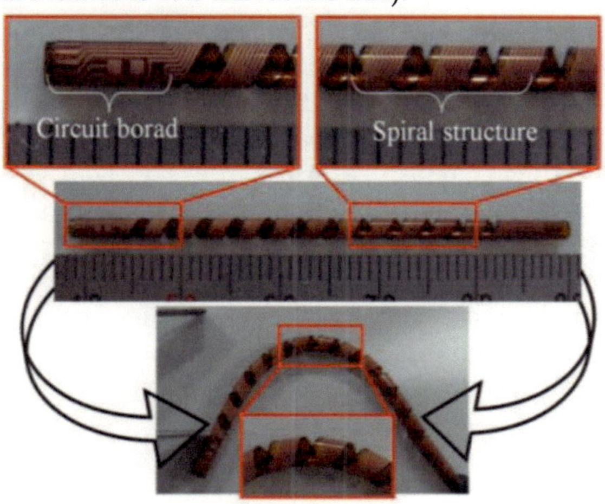

Fig. 13. Fabricated spiral polyimide substrate

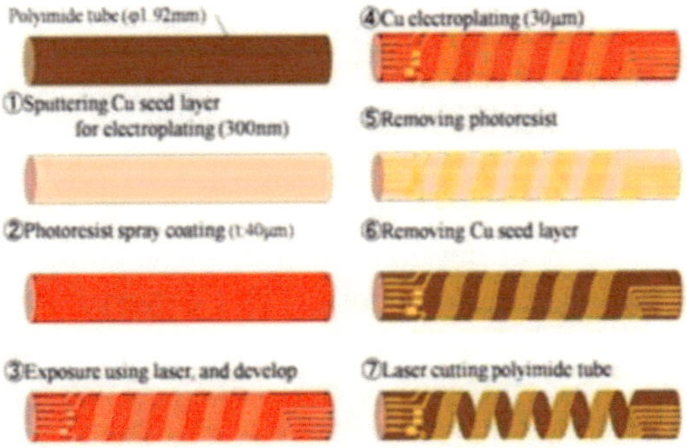

(a) Fabrication process of spiral substrate

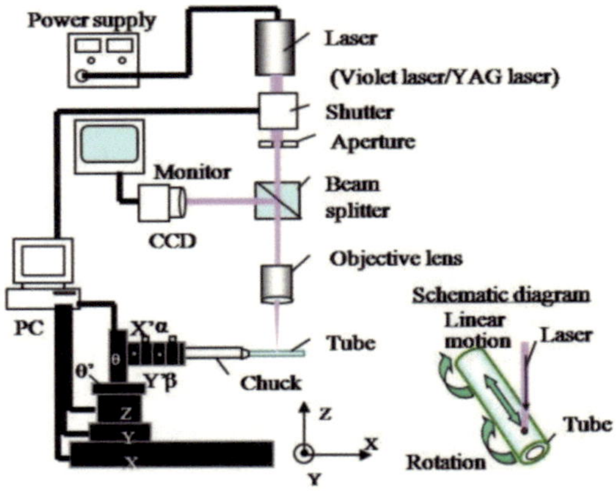

(b) Laser exposure system

Fig. 14. Fabrication process and exposure system

CMOS (Complementary Metal Oxide Semiconductor) imagers have become cheaper and have higher resolution by improvement of the semiconductor processing technology. For colonoscopy, active bending colonoscope with CMOS imager has been fabricated. Circular printed circuit boards were stacked for assembly imaging parts and LEDs as shown in Fig. 17. Fabricated active bending colonoscope is shown in Fig. 17. The external diameter and length of the bending part are 5 mm and 60 mm respectively. The external diameter and thickness of the CMOS imager part are 9 mm and 5 mm. Bending motion is controlled using controller as shown in Fig. 18. The maximum bending angle of the fabricated endoscope is 90 degrees (Curvature radius is 38 mm at 150 mA). Bending motion and inspection using CMOS imager were demonstrated in digestive tract model.

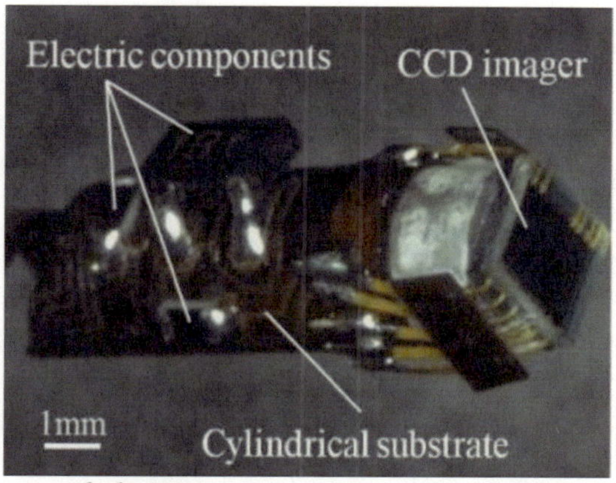

(a) Mounted CCD imager and electric components at a tip of substrate

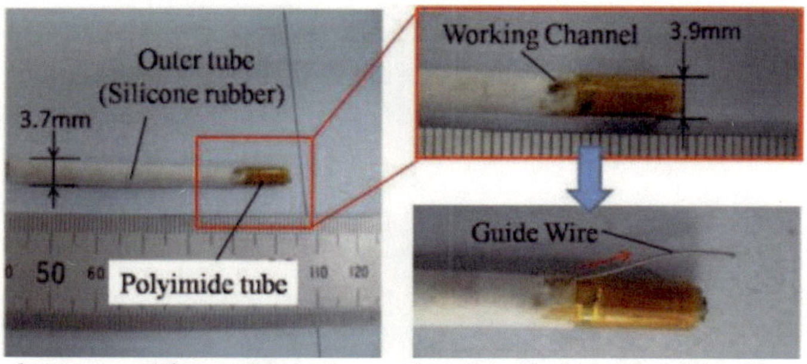

(b) Fabricated endoscope with working channel and insertion of guide wire

Fig. 15. Fabricated small diameter active bending endoscope

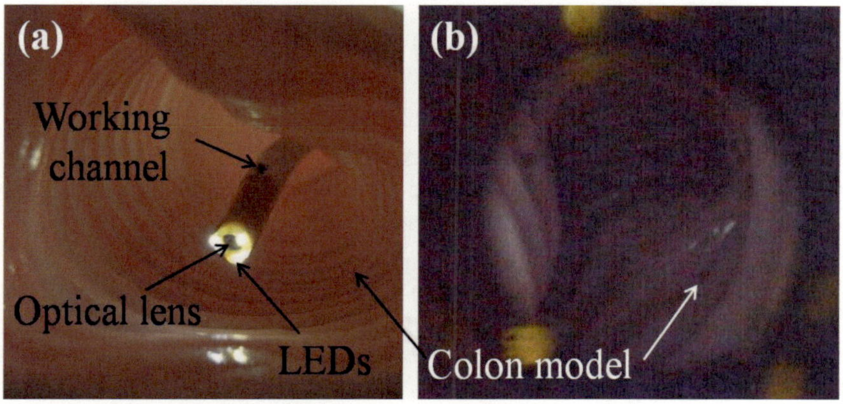

Fig. 16. Motion and inspection in digestive tract model
(a) Inserted endoscope into a digestive tract model
(b) Photograph of digestive tract model using CCD imager

Structure of a bending mechanism using SMA actuator is more simple than that using wires traction. Low cost disposable gastroscopes and colonoscopes can be realized using active steerable mechanism using SMA micro actuators.

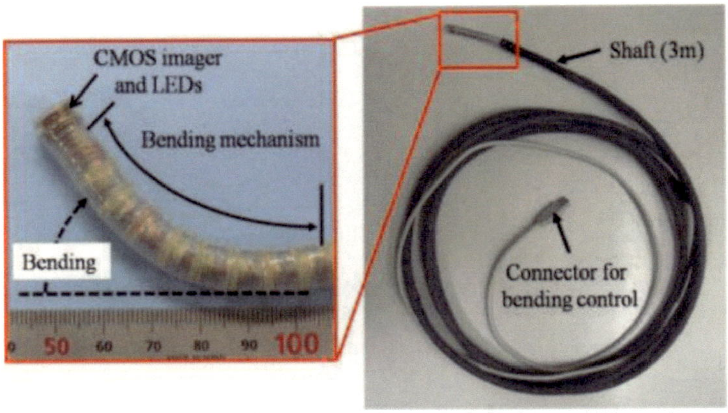

Fig. 17. Fabricated active bending colonoscope (external diameter, 10 mm)

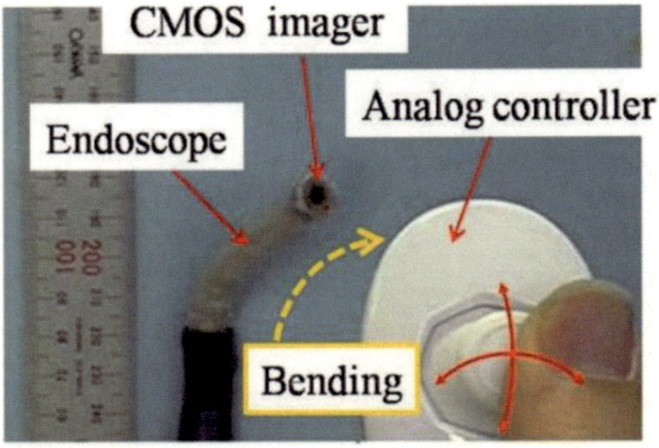

Fig. 18. Controller and bending actuation

6. Conclusion

SMA materials, particularly TiNi-based alloys, have excellent mechanical properties for actuators, such as large force generation and large deformation. Technical progress has resulted in the development of many SMA micro devices.

Medical applications of SMA micro actuator, namely, catheters and endoscopes are mentioned. Active catheters which have a bending mechanism using SMA wires, plates, spiral coil and meandering coil were described. And an active torsional, stiffness control mechanism and a bending mechanism for an active guide wire were described. Active bending endoscopes which have imager at the tip of bending mechanism were also mentioned.

SMA micro actuators have some disadvantages such as low displacement controllability and heat generation. However, thin, soft and low cost medical steering mechanism will effective for realization of active catheters and active endoscopes which enable easy insertion, inspection and precise treatment in the human body.

7. References

Allen, D. M. & Chen, T. T. (1997). The Electrochemical micromachining of microactuator devices from sputterd NiTi thin film, *Electrochemical Society Proceedings*, pp. 72-78, May, 1997

Ballandras, S.; Calin, M.; Zissi, S.; Bertsch, A.; Andre, J. C.; Hauden D. (1997). MicroStereophotolithography and shape memory alloy for the fabrication of miniaturized actuators. *Sensors and Actuators A*, Vol. 62, pp. 741-747

Fukuda, T.; Guo, S.; K, Kosuge.; Arai, F.; Negoro, M. & Nakabayashi, K. (1994). Micro Active Catheter System with Multi Degrees of Freedom, *Proceedings of 1994 IEEE*, International Conference on Robotics and Automation, pp. 2290-2295, San Diego, California, USA, 1994

Hachisuka, K. & Isogai, M. (1991). Indication and limitation of conservative therapy on intestinal obstruction *Gastro-enterological Surgery*, 14, pp. 1639-1645 (in Japanese)

Haga, Y.; Maeda, S. & Esashi, M. (2000). Bending, torsional and extending active catheter assembled using electroplating, *Proceedings of IEEE, International Micro Electro Mechanical Systems Conference (MEMS 2000)*, pp. 181-186, Miyazaki, Japan, January 2000

Haga, Y.; Mizushima, M.; Matsunaga, T. & Esashi, M. (2005). Medical and Welfare Applications of Shape Memory Alloy Microcoil Actuators. *Smart Materials and Structures*, Vol. 14, No. 5, pp. S266-S272

Haga, Y.; Tanahashi, Y. & Esashi, M. (1998). Small diameter active catheter using shape memory alloy, *MEMS 98 Proceedings, IEEE Int. Workshop on Micro Electro Mechanical Systems*, pp. 419-424, 0-7803-4412-X, January 1998, Heidelberg , Germany

Ikuta, K.; Tsukamoto, M. & Hirose, S. (1998). Shape memory alloy servo actuator system with electric resistance feedback and application for active endoscope, *Proceedings of the IEEE Int. Conf. Robotics and Automation*, pp. 427-430, 0-7803-4300-X, Leuven, Belgium, May 1998

Kaneko, S.; Aramaki, S.; Arai, K.; Takahashi, Y.; Adachi, H. & Yanagisawa, K. (1995). Multi-Freedom Tube Type Manipulator with SMA Plate, *Proceedings of the International Symposium on Microsystems, Intelligent Materials and Robots*, pp.87-90, Sendai, Japan, September 1995

Kohl, M.; Quandt, E.; SchBler, A.; Trapp, R. & Allen, D. M. (1994). Characterization of NiTi shape memory microdevices produced by microstructuring of etched sheets or sputter deposited films, *Proceedings of 4th New Actuators (Actuator '94)*, pp. 317-320, Bremen, Germany, June 1994

Lim, A. & Lee, S.(1999). Thin Tube-type Bending Actuator Using Planar S-shaped Shape Memory Alloy Springs. *Technical Digest of the 10th International Conference Solid-State Sensors and Actuators (Transducers '99)*, Vol. 2, pp. 1894-1895, 1999, Sendai

Lim, G.; Minami, K.; Yamamoto, K.; Sugihara, M.; Uchiyama, M. & Esashi, M. (1996). Multi-link Active Catheter Snake-like Motion. *Robotica*, Vol. 14, 1996, pp. 499-506

Lim, G.; Park, K.; Sugihara, M.; Minami, K. & Esashi, M. (1996). Future of active catheters, *Sensors and Actuators A*, Vol. 56, 1996, pp. 113-121

Makishi, W.; Matsunaga, T.; Esashi, M. & Haga, Y. (2000). Active Bending Electric Endoscope Using Shape Memory Alloy Coil Actuators. *Trans. IEE of Japan*, Vol.127-E, No. 2, pp. 75-81 (in Japanese 2000).

Menciassi, A.; Park, J. H.; Lee, S.; Gorini, S.; Dario, P. & Park, J. O. (2002). Robotic solutions and mechanisms for a semi-autonomous endoscope, *Proceedings of the 2002 IEEE/RSJ International Conference on Intelligent Robots and System*, Vol. 2, pp. 1379-1384 , Lausane, Switzerland, October 2002

McCoy, W. C. (1985). *U. S. Patent*, No. 4543090

Mineta, T. (2004). Electrochemical Etching of NiTi Shape Memory Alloy sheet using New Electrolyte Solutions, *J. Micromech and Microeng*, 14, 2004, pp. 76-80

Mineta, T.; Haga, Y. & Esashi, M. (2002). Electrochemical Etching of Shape Memory Alloy Using New Electrolyte Solutions, *Proceedings of 15th IEEE International Micro Electro Mechanical Systems Conference (MEMS2002)*, pp. 376-379, 0-7803-7185-2, Las Vegas, Nevada, USA, January 2002

Mineta, T.; Mitsui, T.; Watanabe, Y.; Kobayashi, S.; Haga, Y. & Esashi, M. (2000). Batch Fabricated Flat Meandering Shape Memory Alloy Actuator for Active Catheter, *Sensors and Actuators A*, 88, pp. 112-120

Mineta, T.; Mitsui, T.; Watanabe, Y.; Kobayashi, S.; Haga, Y. & M. Esashi. (2002). An Active Guide Wire With Shape Memory Alloy Bending Actuator Fabricated by Room Temperature Process. *Sensors and Actuators A*, Vol.97-98C, pp. 632-637

Mineta, T.; Watanabe, Y.; Kobayashi, S.; Haga, Y. & Esashi, M. (1999). Active Catheter Using Multi-Link-Joint Structure Fabricated in Silicon Wafer Trans, *IEE of Japan*, Vol.119-E, No.12, 1999, pp. 615-619

Mizushima, M.; Haga, Y.; Totsu, K. and Esashi, M. (2004). 2004 Active bending long intestinal tube using shape memory alloy, *Proceeding of the Asia-Pacific Conference of Transducers and Micro-Nano Technology (APCOT MNT)*, pp. 893-896

Otsuka, K. & Wayman, C. M. (1998). *Shape Memory Materials*, Cambridge University Press

Park, K.; Minami, K. & Esashi, M. (1996). An integrated communication and control system for a multi-link active catheter. *J. Micromech. Microeng.*, Vol. 6, 1996, pp. 345-351

Reynaerts, D.; Peirs, J. & Brussel, H. V. (1999). Shape memory micro-actuation for a gastro-intestinal intervention system. *Sensors and Actuators A*, Vol. 77, No. 2, 1999, pp. 157-166

Takizawa, H.; Tosaka, H.; Ohta, R.; Kaneko, S. & Ueda, Y. (1999). Development of a Microfine Active Bending Catheter Equipped with MIF Tactile Sensors, *Twelfth IEEE International Conference on Micro Electro Mechanical Systems (MEMS '99)*, pp. 412-417, 1999, Orlando

7

Numerical simulation of a semi-active vibration control device based on superelastic shape memory alloy wires

Corneliu Cismaşiu and Filipe P. Amarante dos Santos
Centro de Investigação em Estruturas e Construção - UNIC, Faculdade de Ciências e Tecnologia, Universidade Nova de Lisboa, 2829-516 Caparica Portugal

1. Introduction

The martensitic transformations exhibited by shape memory alloys (SMAs) are diffusionless phase transformations in solids, that change the material from the parent phase, a cubic crystal structure called austenite, to a lower symmetry monoclinic crystal structure, called martensite. These phase transformations yield the shape memory effect and superelasticity, peculiar thermo-mechanical properties which make the SMAs very appealing for applications in Civil Engineering structures.

Superelasticity allows the material to recover from large nonlinear strains during a mechanical cycle of loading and unloading, while dissipating a considerable amount of energy through hysteresis. The chemical composition of the alloy, as well as its thermomechanical treatment, has an important influence on the SMA properties and on the extent of the area enclosed by the hysteretic loop (Ortín & Delaey, 2002; Otsuka & Wayman, 1998).

Exploiting this property, several authors report on successful implementations of SMAs in passive vibration control devices. Most of them are especially designed for seismic hazard mitigation. They range from simple austenitic superelastic (SE) wires in bracing systems (Torra et al., 2007) to hinge restrainers SE bars in bridges (DesRoches & Delemont, 2002). A combination of martensitic bars or austenitic SE wires with pre-stressed SE wires working in phase opposition have also been successfully implemented both in bracing and isolation systems (Dolce et al., 2000; 2005). Additionally, steel beam to column connections, built up of martensitic bars, have also been reported (Ocel et al., 2004). Most of these devices are built up of Nitinol (NiTi), a dual alloy built up of Nickel and Titanium (Zhang & Zu, 2007), due to its outstanding fatigue properties, high corrosion resistance and ductility. A hybrid base isolation system, composed of linear elastomeric bearings, friction-pendulum bearings, SE austenitic wires and magnetorheological dampers was proposed by Shook *et al.* (Shook et al., 2008) for mitigation of seismic motions. To manage the superstructure response to ground motions, each subcomponent of the isolation system was designed for a specific task. The SE austenitic wires were used to supply recoverable hysteretic behavior and to serve as an additional restoring force. Zhang and Zu (Zhang & Zu, 2007) propose an adaptive vibration control device based on SE austenitic wires that presents two distinctive features: its tunable hysteretic behavior and the ability to withstand several design level earthquakes. The hysteretic behavior of the damper could be modified to best fit passive structural control applications by adjusting dampers design parameters such as the inclination angle of the SE wires, pre-tension level and the friction coefficient.

The vibration control device proposed in the present work, originates from the NiTi based passive dissipation device reported by Dolce *et al.* (Dolce et al., 2000). As to increase its energy dissipation capabilities, the original device was made of sets of pretensioned wires working in phase opposition. However, a variety of relaxation phenomena observed in pretensioned SE wires (Bhattacharyya et al., 2002; Matsui et al., 2004) might have an adverse impact on the dynamic performances of the device. Mainly to avoid these problems, the proposed semi-active device uses a strategy that allows the continuous adapting of the accumulated strain in the wires, based on the response of the device to external excitations. As in an active control system, a controller monitors the feedback measurements and generates appropriate command signals for the device. As in a passive control system however, the control forces are developed as a result of the motion of the structure itself, with no need of external energy input. As the control forces act as to oppose the motion of the structural system, they promote the global stability of the structure (Symans & Constantinou, 1999). According to presently accepted definitions (Housner et al., 1997), the proposed device may be considered as a semi-active control system.

2. Thermo-mechanical properties

While some other commercial forms are also available, namely tubes, strips and sheets, NiTi is most commonly used in form of wire or bars. In order to characterize it's tensile behavior in wire form, three sets of experimental tests were perform on NiTi SE508 wire specimens, with 2.4 mm diameter and circular cross section, provided by Euroflex GmbH.

2.1 DSC thermo-analytical analysis

In the stress-free state, a SMA is characterized by four transformation temperatures: M_s and M_f during cooling and A_s and A_f during heating. The first two (with $M_s > M_f$) indicate the temperatures at which the forward transformation starts and finishes, respectively. The last two (with $A_s < A_f$) are the starting and finishing temperatures for the inverse transformation. One of the most used methods to identify these transformation temperatures, is the Differential Scanning Calorimetry (DSC) (Höhne et al., 2003), a thermo-analytical technique in which the difference in the amount of heat required to increase the temperature of a sample and a reference of well defined heat capacity, are measured as a function of temperature. Both the sample and the reference are maintained at nearly the same temperature throughout the experiment. The basic principle underlying this technique is that, when the sample undergoes physical transformations such as phase transitions, a different quantity of heat is needed to maintain it at the same temperature as the reference. If it's more or less, depends on whether the process is exothermic, as the forward martensitic transformation, or endothermic, as the inverse martensitic transformation. Care must be taken as the presence of an eventual R-phase prior to thermal cycling and residual stresses from sample cutting tends to complicate the curves and introduce spurious peaks (Collings, 1995). In order to characterize the transformation temperatures of the SE NiTi SE508 wire specimen, a DSC test was performed using a SETARAM-DSC92 thermal analyzer, Figure 1(a). Although the sample was tested as received, prior to the DSC analysis it was submitted to a chemical etching in order to remove the oxide and the layer deformed by the cutting operation. During the DSC analysis, the sample was heated up to 80°C, held at this temperature for 6 min and then cooled to -80°C, with heating/cooling rates of 7.5°C/min. A baseline analysis of the resulting curves is presented in Figure 1(b), where the four transformation temperatures can be easily identified. Their values are presented in Table 1.

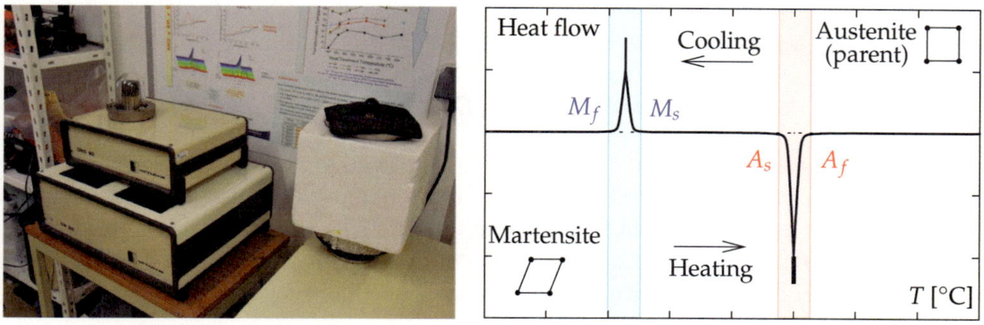

(a) SETARAM-DSC92 thermal analyzer (b) Baseline analysis of the resulting DSC curves

Fig. 1. Differential Scanning Calorimetry thermo-analytical analysis

2.2 Quasi-static tensile tests at room temperature

To complete the material characterization, the SE NiTi SE508 wire specimen was also subjected to a pseudo-static, strain driven uniaxial tensile test on a Zwick/Roell Z050 testing machine, Figure 2(a), with a strain rate of 0.02%/s. The tests were conducted at a room temperature of 24°C up to a maximum strain of 6%.

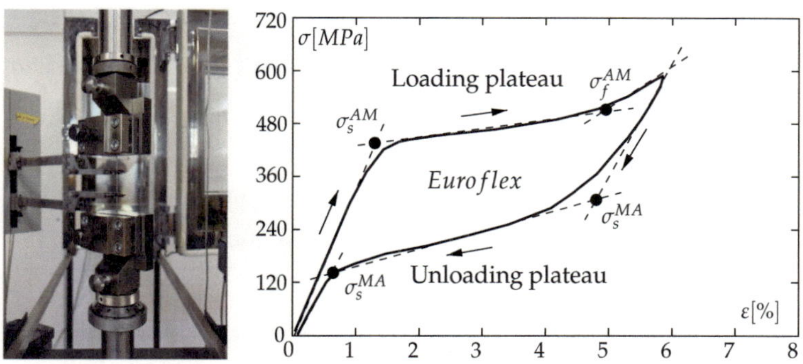

(a) Zwick/Roell Z050. Gripping apparatus (b) Stress-strain response of NiTi wire

Fig. 2. Quasi-static uniaxial tensile test

The usual shape of a superelastic hysteretic cycle, characterized by two regions of almost constant stress, the loading and the unloading plateaux during the forward and inverse transformations, is clearly captured, as illustrated in Figure 2(b). Based on the inspection of this resulting stress-strain response, the austenitic and martensitic elastic modulus (Auricchio & Sacco, 1997), the maximum residual strain, as well as the starting and final stresses during the forward and inverse transformation, are identified and collected in Table 1. For the Poisson's ratio, a typical value of 0.33 is adopted from the literature (Collings, 1995).

2.3 Quasi-static tensile tests at different ambient temperatures

Another parameter needed in the definition of the constitutive model is the Clausis-Clapeyron coefficient (CCC), representing the slope ($d\sigma/dT$) of the lines defining the boundaries of the SMA transformation strips in the material's phase plane diagram, Figure 6(a). Note that, the CCC is assumed to be the same for the forward, C_M and inverse, C_A, transformations.

According to (Isalgue et al., 2008), the most straightforward experimental procedure to compute the CCC of a previously stabilized SMA specimen is to determine it's critical transformation stresses for a set of different ambient temperatures. Based on this principle, an experimental testing procedure was implemented and conducted using a Zwick/Roell Z050 testing machine and a W91255 temperature controlled chamber, Figure 3.

(a) (b) (c)

Fig. 3. W91255 temperature controlled chamber (a) prior and (b) during the temperature controlled tensile testing and (c) detail of the thermocouple

The SE NiTi SE508 wire specimen was strained up to 7.5% with a strain rate of 0.068%/s, in an ambient temperature that ranged from 30 to 78°C, with an increment of 3°C. The temperature in the wire was constantly monitored by a thermocouple placed at it's mid section.

The resulting stress-strain curves are displayed in Figure 4(a). Observing these graphs, it is clear that the superelastic hysteresis shifts upwards as the ambient temperature rises. In order to estimate the CCC, the stress variation with temperature must be evaluated. As particular hysteretic points, like, for instance, the one associated with the beginning of the forward transformation, may be difficult to identify, the average variation of the stress on the loading plateau was used. For the prescribed temperature increment, its value was 19.5 MPa, leading to a CCC of 6.5 $MPaK^{-1}$. Experimental results reported in the literature (Isalgue et al., 2008) present similar CCC values, 6.6±0.2 $MPaK^{-1}$, in a NiTi wire specimen with 2.46 mm diameter. The plot in Figure 4(b), represent the linear relation between temperature and the stress

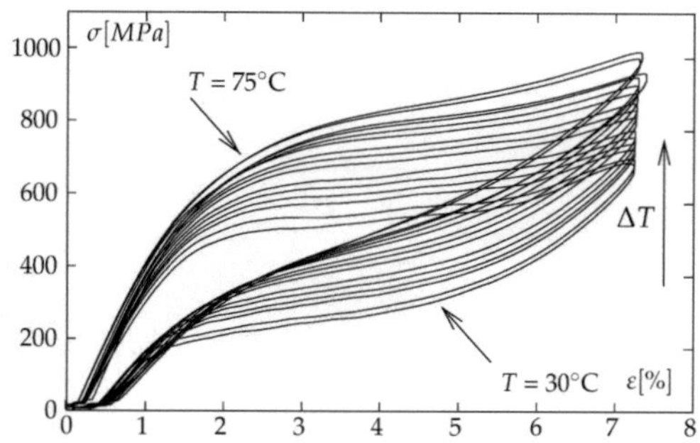

(a) Progressive stress induced transformation cycles

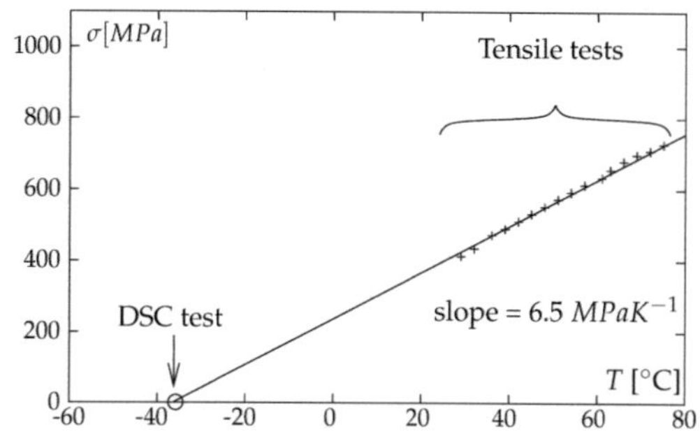

(b) Stress-temperature relation

Fig. 4. Temperature dependence of the stress induced martensitic transformation in quasi-static uniaxial tensile tests

associated with the beginning of the forward transformation, and it was obtained using the estimated for the CCC and the M_s value from the DSC analysis.

3. Constitutive model

The behavior of SMAs denotes a high level of complexity, as it depends on stress and temperature and is closely connected with the crystallographic phase of the material and the thermodynamics underlying the transformation processes. For this reason, any simulation regarding the design of a particular system that exploit the interesting features exhibited by these materials, must be based on an adequate constitutive model. The model must be able to conveniently cope with the relevant set of SMA's properties regarding the type of application at stake. While strain or stress rate independent models (Auricchio et al., 2006; 1997; McCormick et al., 2006), may be well suited for quasi-static analysis when isothermal conditions may be assumed, for most dynamic applications, rate dependent models are usually required, due to self-heating (Ivshin & Pence, 1994; Liang & Rogers, 1990; Tanaka et al., 1986).

Several constitutive mathematical models are currently available in the literature, most of them being aimed at the one-dimensional description of the material behavior (Auricchio & Sacco, 1997; Brinson & Huang, 1996; Liang & Rogers, 1990). One of the common features present in most of them, is the presence of a distinct mechanical law, governing the stress-strain relations, and of a kinetic law, governing the martensitic transformations. In the case of rate dependent models, to conveniently model the thermal effects, a heat balance equation is coupled with the mechanical and kinetic transformation laws.

3.1 Mechanical laws

In a typical SMA constitutive model, the mechanical law relates the stress (σ), strain (ε), temperature (T) and martensite fraction (ξ). The martensite fraction is an internal state variable, representing the extent of the transformation in the material and can be regarded as the fraction of the produced phase. This transformed phase fraction is considered to be in series with the elastic fraction. According to the literature (Brinson & Huang, 1996), there are several possible approaches to model the elastic component. If considered to be limited to the austenitic

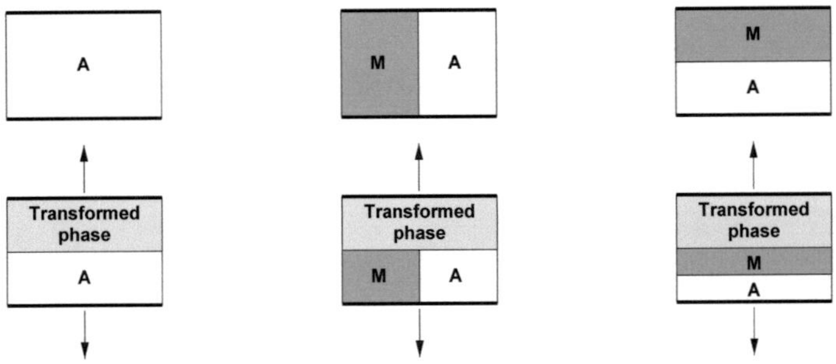

(a) Simple serial model (b) Voight scheme (c) Reuss scheme

Fig. 5. SMA's mechanical laws

phase, the simple serial model presented in Figure 5(a) is obtained. If, however, two crystallographic phases, the austenite and martensite[1] are to be contemplated in this elastic phase, two cases can be considered, namely the Voight and the Reuss schemes. The Voight scheme, Figure 5(b), considers a parallel distribution of austenite and martensite within the material, while the Reuss scheme, Figure 5(c) is based on the assumption of a periodical distribution of austenite and martensite within the material, orthogonal to the direction of the applied stress.

3.1.1 Simple serial model

In the case of the simple serial model illustrated in Figure 5(a), as the principle of superposition applies, the total strain yields adding the elastic $\varepsilon_{elast.}$ to the transformation strain $\varepsilon_{transf.}$,

$$\varepsilon = \varepsilon_{elast.} + \varepsilon_{transf.}$$

Knowing that the transformation strain can be expressed function of the maximum residual strain in the material , ε_L, and the martensite fraction as $\varepsilon_L \xi$, and taking into account that the elastic strain equals the austenitic strain, ε_A, the total strain reads

$$\varepsilon = \varepsilon_A + \varepsilon_L \xi \qquad \text{with} \quad 0 \leq \xi \leq 1$$

and the mechanical law relating the strains and the stresses yields

$$\sigma = \sigma_A = E_A \varepsilon_A = E_A(\varepsilon - \varepsilon_L \xi)$$

where E_A is the austenitic elastic modulus. The model neglects the additional thermal strain component, which is several orders of magnitude smaller than the transformation strain (Brinson & Huang, 1996).

3.1.2 Voight scheme

In the Voight scheme, as clearly illustrated in Figure 5(b), the elastic, austenitic and martensitic strains are all equal and therefore, the mechanical law governing the material behavior can be written as

$$\sigma = \xi \sigma_M + (1 - \xi)\sigma_A$$

[1] Not distinguishing between the twinned and the detwinned martensite.

Using the definition for the austenitic and martensitic stresses, $\sigma_A = E_A\varepsilon_A$ and $\sigma_M = E_M\varepsilon_M$, the mechanical law can be expressed as

$$\sigma = [\xi E_M + (1-\xi)E_A]\,(\varepsilon - \varepsilon_L\xi)$$

where ε_M and E_M are the martensitic strain and elastic modulus, respectively. Adding a thermal strain component to the above result, yields the mechanical law proposed by Tanaka (Tanaka et al., 1986),

$$\sigma = [\xi E_M + (1-\xi)E_A]\,(\varepsilon - \varepsilon_L\xi) + \theta(T - T_0)$$

where θ is the thermal coefficient of expansion and T_0 the temperature at which the thermal strain is defined to be zero.

3.1.3 Reuss scheme

In the Reuss scheme, and according to Figure 5(c), it can be seen that the total strain can be expressed as

$$\varepsilon = \xi\varepsilon_M + (1-\xi)\varepsilon_A$$

Taking into account that the stress in each phase is equal to the overall applied stress and using the definition of the austenitic and martensitic strains, $\varepsilon_A = \sigma/E_A$ and $\varepsilon_M = \sigma/E_M + \varepsilon_L$, the mechanical law can be expressed as

$$\sigma = \frac{E_A E_M}{\xi E_A + (1-\xi)E_M}(\varepsilon - \varepsilon_L\xi)$$

The Ivshin-Pence mechanical law (Ivshin & Pence, 1994) can be recovered by adding the thermal strain component to the above equation, to read,

$$\sigma = \frac{E_A E_M}{\xi E_A + (1-\xi)E_M}(\varepsilon - \varepsilon_L\xi) + \theta(T - T_0)$$

3.2 Kinetic laws

The transformation kinetic laws describe mathematically the evolution of the martensite fraction with stress and temperature, based on the material's stress-temperature phase diagram. Several kinetic laws are available in the literature, ranging from linear (Auricchio & Sacco, 1997) or exponential laws (Tanaka et al., 1986), to cosine based kinetic laws (Liang & Rogers, 1990) and thermodynamically derived relations (Ivshin & Pence, 1994).

The linear and the exponential kinetic laws are briefly described in what follows.

3.2.1 Linear transformation kinetic law

The linear transformation kinetic laws are based on the assumption that the relation between strains and stresses are represented by a series of linear segments. The material constants, M_f, M_s, A_s, A_f, C_M and C_A, are illustrated in the stress-temperature phase diagram in Figure 6(a), where an isothermal, stress induced martensitic transformation path is drawn. The starting and final stresses during the forward, σ_s^{AM} and σ_f^{AM}, and inverse transformations, σ_s^{MA} and σ_f^{MA}, are represented in the corresponding strain-stress diagram. As the scalar variable ξ represents the martensite fraction during the phase transformation, its variation occurs during the conversion of austenite into martensite ($A \rightarrow M$) or of martensite into austenite ($M \rightarrow A$),

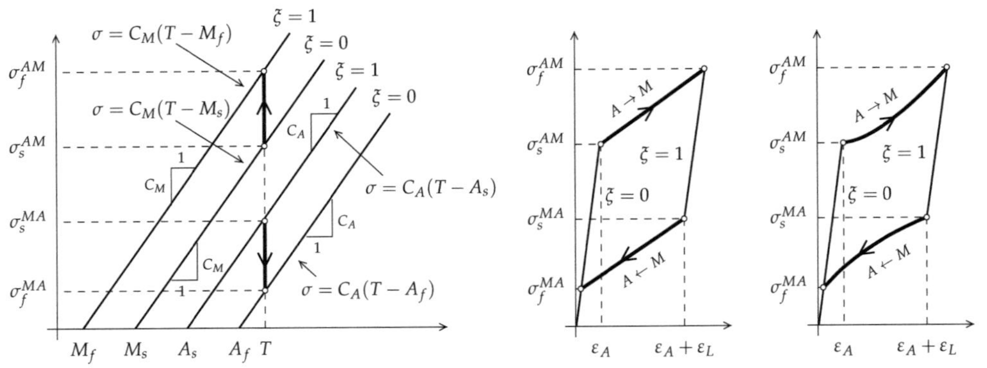

(a) Stress-temperature phase diagram (b) Linear kinetics (c) Exponential kinetics

Fig. 6. Phase diagram and transformation kinetics

the forward and inverse transformations, respectively. In the case of linear kinetics, the function $\xi(\sigma)$ presents a linear behavior, assuming values between zero and one, corresponding to the pure austenite and martensite phases. Hence, for a specific temperature, T, the kinetic laws describing the forward and inverse transformations can be expressed as

$$\xi^{AM} = \frac{\sigma - \sigma_s^{AM}}{\sigma_f^{AM} - \sigma_s^{AM}} \qquad \xi^{MA} = \frac{\sigma - \sigma_f^{MA}}{\sigma_s^{MA} - \sigma_f^{MA}}$$

The equations defining the boundary lines of the transformation strips represented in Figure 6(a) are introduced in the equations above to yield the linear kinetic laws for the forward and inverse transformations, respectively,

$$\xi^{AM} = \frac{\sigma - C_M(T - M_s)}{C_M(M_s - M_f)} \qquad \xi^{MA} = \frac{\sigma - C_A(T - A_f)}{C_A(A_f - A_s)}$$

3.2.2 Exponential transformation kinetic law

The exponential kinetic laws are based on the Magee's equations, originally proposed for steels (Magee, 1970). For the forward and inverse transformations, they yield,

$$\frac{\dot{\xi}^{AM}}{1 - \xi^{AM}} = a_M \dot{T} - b_M \dot{\sigma} \qquad -\frac{\dot{\xi}^{MA}}{\xi^{MA}} = a_A \dot{T} - b_A \dot{\sigma}$$

The well known (Vitiello et al., 2005) exponential kinetic laws results by integrating the above equations from the initial values (σ_0, T_0, ξ_0) to the current values (σ, T, ξ), for constant material parameters a_M, b_M, a_A and b_A and taking into account that $b_M C_M = a_M$ and $b_A C_A = a_A$, to yield, for the forward and inverse transformations,

$$\xi^{AM} = 1 - \exp\left[a_M(M_s - T) + b_M\sigma\right] \qquad \xi^{MA} = \exp\left[a_A(A_s - T) + b_A\sigma\right]$$

To determine the material constants a_M and b_M, one follows the usual understanding in metallurgy (Jones & Ashby, 1998), assuming that the forward transformation is complete when

$\xi = 0.99$. The material constants a_A and b_A are calculated in a similar way, assuming that the inverse transformation is complete when $\xi = 0.01$. Using these assumptions, simple algebraic manipulations yield,

$$a_M = -\frac{2\ln 10}{M_s - M_f} \qquad b_M = \frac{a_M}{C_M} \qquad a_A = \frac{2\ln 10}{A_f - A_s} \qquad b_A = \frac{a_A}{C_A}$$

3.3 Heat balance equation

When quasi-static loading conditions can be assumed, the heat exchanges between the superelastic material and it's surrounding environment generate almost isothermic processes. However, as the rate of dynamic loading increases, since the dissipation capacity of the thermo-mechanical system is limited by the heat convection mechanism, the generated and the dissipated energy become unbalanced, causing changes in the specimen's temperature and shape of the hysteretic loop, and consequently, in the damping capacity of the material itself. In order to attend this phenomena, whilst modeling the dynamical behavior of these materials, the mechanical and kinetic laws must be coupled with a heat balance equation considering the thermal effects on the material (Cismasiu & Santos, 2008; Koistinen & Marburger, 1959; Vitiello et al., 2005).

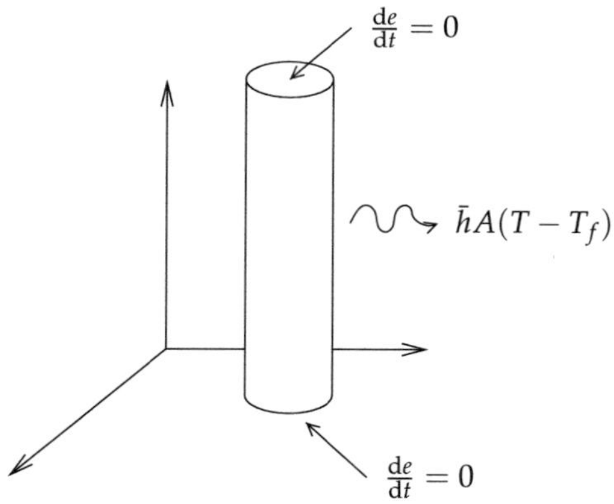

Fig. 7. Heat transfer system

The heat transfer system under analysis, presented in Figure 7, consists of a superelastic NiTi wire with circular cross section, fixed at both ends and surrounded by air, at temperature T_f. There are internal energy sources within the wire, deriving from the enthalpy of the martensitic transformations and internal friction, both occurring during a hysteretic superelastic cycle. It is assumed that the heat conduction through the wire's extremities is negligible and therefore, it is possible to consider only the lateral surface for calculating the heat exchanges with the environment. The conduction heat transfer phenomena is described by the Fourier's law (Holman, 2009),

$$q = -k\nabla T$$

in which the heat flux q is expressed as a function of the temperature profile and the thermal conductivity, k. The minus sign ensure that the heat flows downwards the temperature gradient. According to the conservation of energy, for a small element within the body, the non-stationary conductive problem can be described by the differential equation of heat conduction with a generating term,

$$\nabla q = q_{gen} - \frac{de}{dt}$$

The power generated per unit of volume, q_{gen}, describes the rate of heat generation within the body. Typically, this new heat must be conducted to the boundaries and removed via convection and/or radiation heat transfer. The change in the internal energy, e, is related to the body's ability to store heat by raising its temperature, and can be expressed as

$$\frac{de}{dt} = \rho c \frac{dT}{dt}$$

where ρ is the density of the material and c it's specific heat. The heat balance equation results substituting the Fourier's law and the definition of the change in the internal energy in the differential equation of the conduction of heat, to yield

$$\nabla^2 T - \frac{\rho c}{k}\frac{dT}{dt} = -\frac{q_{gen}}{k}$$

This equation can be reduced to a simpler form if one can assume a uniform temperature within the specimen, as, under such conditions, the temperature depends only on time. To check the applicability of the above assumption, one can compute the Biot number, Bi, a dimensionless value that gives a simple index of the ratio of the heat transfer resistance inside and at the surface of a body. This ratio determines whether or not the temperature inside a body will vary significantly in space, while the the body heats or cools over time. The Biot number is defined as

$$Bi = \frac{\bar{h} L_c}{k}$$

where L_c is the characteristic length, commonly defined as the volume of the body divided by its surface area, $\bar{h}$ is the mean coefficient of heat transfer and k the thermal conductivity. Values of the Biot number less than 0.1 imply that the heat conduction inside the body is much faster than the heat convection away from its surface and therefore, the temperature gradients are negligible inside of it. In these cases, the use of a one-dimensional analysis leads to errors less than 10% (Holman, 2009) and the heat balance equation can be simplified to,

$$-\rho c V \frac{dT}{dt} = \bar{h} A (T - T_f) - q_{gen} V \qquad \text{with} \quad T(0) = T_f$$

where V is the volume of the body and A its surface area. The power generated per volume unit, appearing in the above equation, is defined as,

$$q_{gen} = c_L \rho \frac{d\xi}{dt} + \frac{dW}{dt}$$

where the first term is related to the martensite fraction, assuming constant latent heat, c_L. The second is related to internal friction. As in a complete tensile loading-unloading cycle, the dissipated energy by internal friction, W, corresponds to the total area enclosed by the hysteretic loop, the associated power results dividing this area by the duration of the cycle.

3.4 Adopted constitutive model

The rate-dependent constitutive model used in the following numerical simulations is based on Tanaka's temperature dependent model (Tanaka et al., 1986). In order to relate stress, strain, temperature and martensite fraction in the material, the model couples the constitutive relations to an exponential kinetic law that describes the volume fraction of austenite (Koistinen & Marburger, 1959). When associated to a balance equation that considers the thermal effects on the material, this constitutive model produces reliable results even for high strain rates (Cismasiu & Santos, 2008; Vitiello et al., 2005).

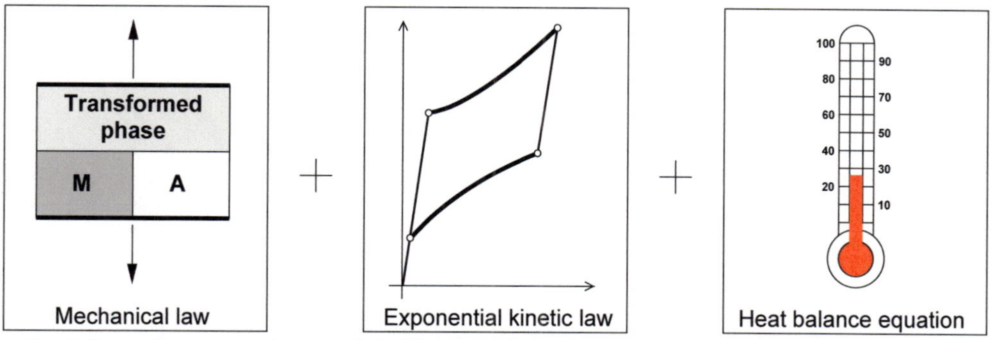

Fig. 8. Adopted constitutive model: Voight scheme, exponential kinetics and heat balance equation

The values reported in (Vitiello et al., 2005) for the following thermophysical properties: latent heat, c_L, specific heat, c_p, density, ρ, convection coefficient, h and thermal expansion coefficient, θ, are used to complete the model characterization. All these parameters are presented in Table 1.

$E_A = 35000$ MPa	$E_M = 20000$ MPa	$M_f = -45°$ C
$M_s = -35°$ C	$A_s = -15°$ C	$A_f = -5°$ C
$C_M = 6.5$ MPaK^{-1}	$C_A = 6.5$ MPaK^{-1}	$e_L = 3\%$
$\rho = 6500$ kg m^{-3}	$c_L = 12914$ J kg^{-1}	$c_p = 500$ J kg^{-1}K^{-1}
$h = 35$ W m^{-2} K^{-1}	$\theta = 6 \times 10^{-6}$ K^{-1}	$k = 18$ W m^{-2} K^{-1}

Table 1. Parameters for adopted constitutive model

3.5 Numerical assessment of the adopted constitutive model

The results of a large set of experimental cyclic tensile tests on 2.4 mm diameter NiTi S wires provided by *Euroflex GmbH*, clearly illustrating the sensitivity of the mechanical response to temperature and strain rate, are used to assess the performances of the constitutive model. All tests were conducted on a Zwick/Roell Z050 testing machine, equipped with a W91255 temperature controlled chamber, Figures 2(a) and 3.

3.5.1 Quasi-static tensile tests

To start with, the experimental curves obtained from a set of quasi-static tensile tests, carried out at a strain rate of 0.020%/s in an ambient temperature of 24°C, were compared with the numerical model estimates. As is illustrated in Figure 9(a) and 9(b), for the case of three complete cycles of decreasing amplitude and for the case of a complete cycle followed by a smaller inner cycle, the model is able to describe a wide variety of loading histories, with numerical predictions that fit well the experimental data. Particularly good estimates are obtained during the forward transformation, when the relative error between the experimental data and the numerical simulation is practically negligible. However, upon unloading, apart from the initial segment, where the stiffness is accurately captures, the model seems to overestimate the SMA's stiffness. This can be explained by the fact that the constitutive model does not capture the gradual transition to the inverse transformation plateau.

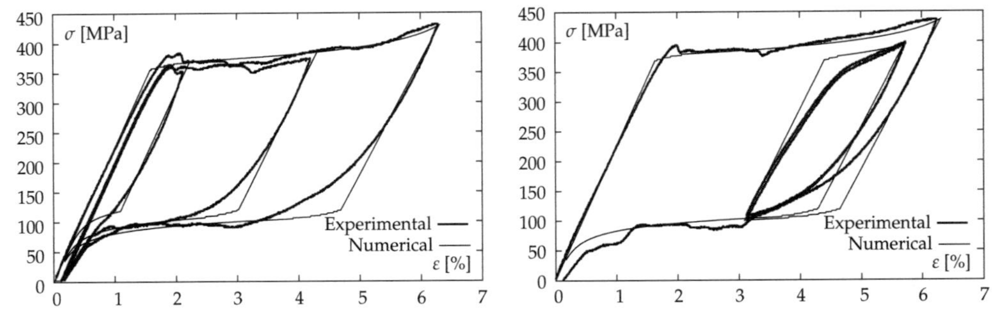

(a) Cycles of decreasing amplitude (b) Cycle followed by a smaller inner cycle

Fig. 9. Quasi-static experimental tests: numerical model versus experimental data

3.5.2 Dynamic tensile tests

The ability of the adopted constitutive model to conveniently model the complex thermo-mechanical response of the superelastic NiTi wire during dynamic loading is investigated next. The experimental curves obtained in four tensile tests, carried out at different strain rates, $\dot{\varepsilon} = 0.008, 0.067, 0.250$ and 0.333%/s, in an ambient temperature of 20°C, are compared with the simulated estimates. The plots in Figure 10 clearly illustrate that the implemented rate-dependent numerical model yields a set of very satisfying results, both for the temperature time histories and their corresponding stress-strain diagrams. It can be seen that, as the strain rate of the dynamic loading increases from 0.008 to 0.333%/s, the simulation predicts increasing changes in the sample temperature during the loading cycle, in close agreement with the experimental data. In what concern the stress-strain diagram, the shape of the hysteretic loops tends to become steeper and narrower with the increase in the strain rate, again in conformity with the trend observed in the experimental tests.

4. Proposed semi-active vibration control device under harmonic excitation

The proposed semi-active control device, whose generic functional scheme is given in Figure 11(b), is based on the passive system illustrated in Figure 11(a), being built up of two SE wires working in phase opposition. In passive systems of this type, the wires are connected to

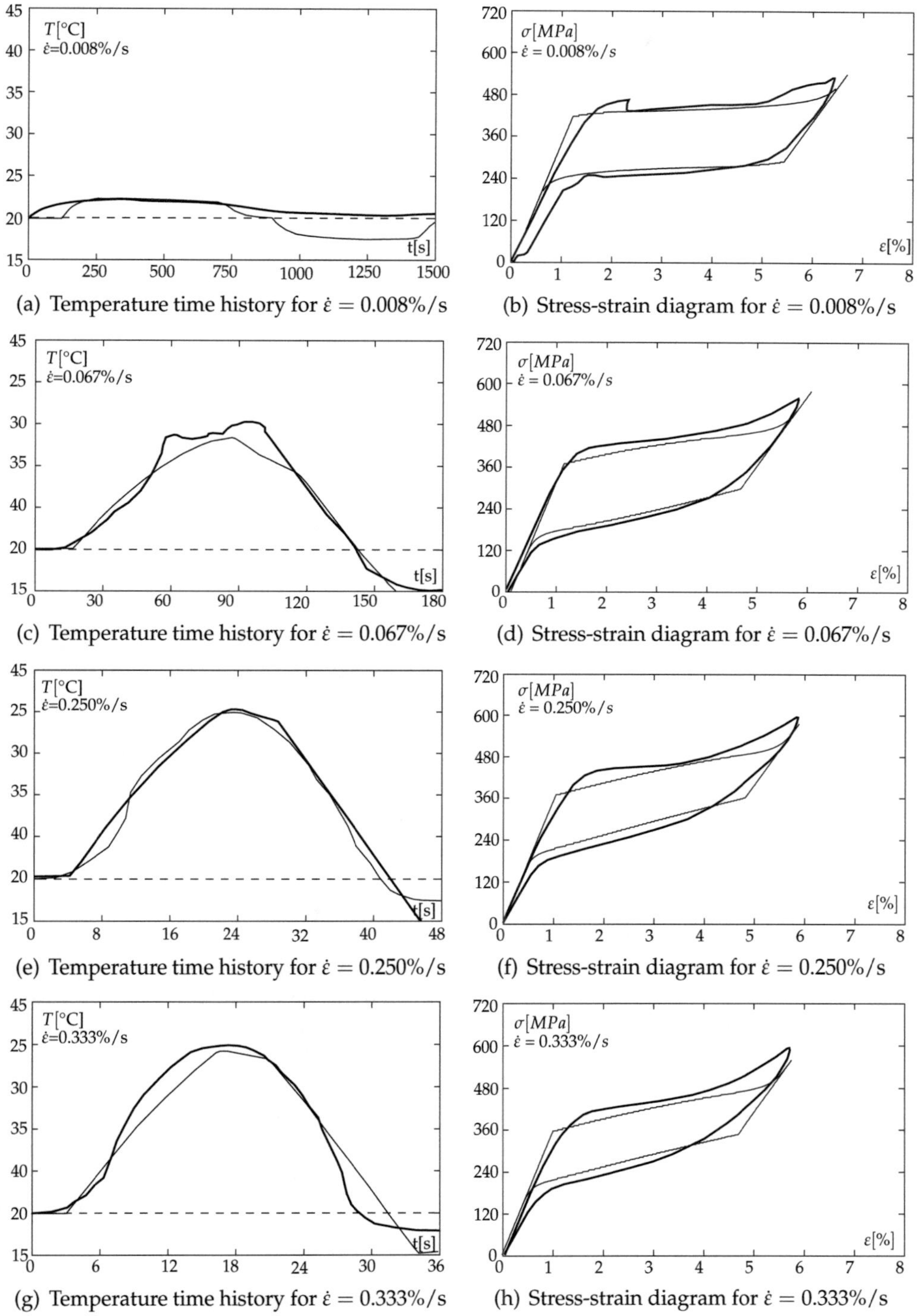

(a) Temperature time history for $\dot{\varepsilon} = 0.008\%/s$

(b) Stress-strain diagram for $\dot{\varepsilon} = 0.008\%/s$

(c) Temperature time history for $\dot{\varepsilon} = 0.067\%/s$

(d) Stress-strain diagram for $\dot{\varepsilon} = 0.067\%/s$

(e) Temperature time history for $\dot{\varepsilon} = 0.250\%/s$

(f) Stress-strain diagram for $\dot{\varepsilon} = 0.250\%/s$

(g) Temperature time history for $\dot{\varepsilon} = 0.333\%/s$

(h) Stress-strain diagram for $\dot{\varepsilon} = 0.333\%/s$

Fig. 10. Dynamic experimental tests at ambient temperature of 20°C: numerical model versus experimental data the mass, m, and act as restoring elements. During the loading cycles, while one of the wires is loaded the other one is unloaded and conversely (Cismasiu & Santos, 2008; Dolce et al., 2000; 2005).

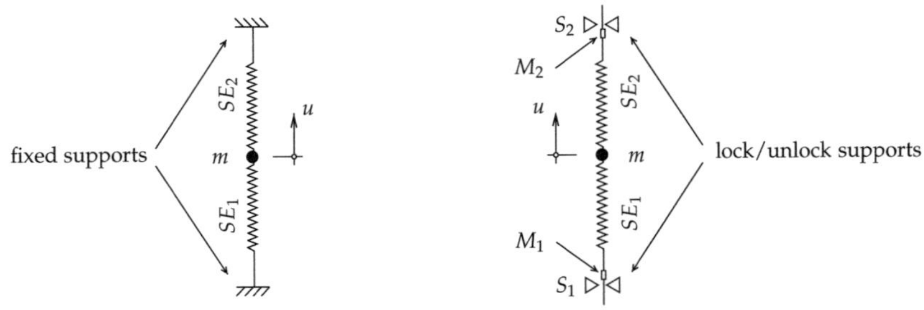

(a) Passive system (b) Semi-active system

Fig. 11. Generic functional scheme of the passive and the semi-active systems: superelastic wires (SE_1 and SE_2), supports (S_1 and S_2) and markers (M_1 and M_2)

The response of a passive vibration system, with and without pre-strain, to harmonic ground accelerations is recalled, to emphasize their dynamic characteristics. Their drawbacks are identified and used as conditioning design parameters for the proposed semi-active system. In the numerical simulations that follows, the system, with zero initial conditions, is excited during 1 second by a 2 Hz harmonic ground acceleration with an amplitude of 1.25 ms^{-2}. The analyses consider also a sequential phase of free oscillations. In all these numerical tests, the nonlinear equation of motion is solved using a Newmark predictor-corrector implicit algorithm (Chopra, 2001), implemented in a specially designed numerical framework (Cismasiu & Santos, 2008).

4.1 Passive system with no pre-strain

The generic functional scheme of such a passive system, with the two superelastic elements, SE_1 and SE_2, each one composed by 10 NiTi wires with a length of 1000 mm and 0.24 mm diameter, is given in Figure 11(a). The mass of the system is 100 kg and the SMA's material parameters are the ones in Table 1. Its strain and displacement time history responses to the prescribed harmonic excitation are presented in Figure 12. Facing compression, the wires exhibit zero stiffness and therefore SE_1 and SE_2 work alternately, as can be observed in Figure 12(a). In this particular case, the extent of the martensitic transformation caused by the dynamic displacements of the mass is quite small, corresponding to a martensite transformation ratio of about 15 % and, hence, little damping occurs. The beginning of the martensitic transformation is represented in Figure 12(a) by the horizontal dot line, $\varepsilon_{Ms} \simeq 1.25\%$.

4.2 Passive system with pre-strain

A common technique to enhance the damping capability of the system is to enforce a pre-strain in the SE wires (Dolce et al., 2000). For a given dynamic action, the value of the pre-strain can be computed as to guarantee that the wires will always be in tension and therefore, one

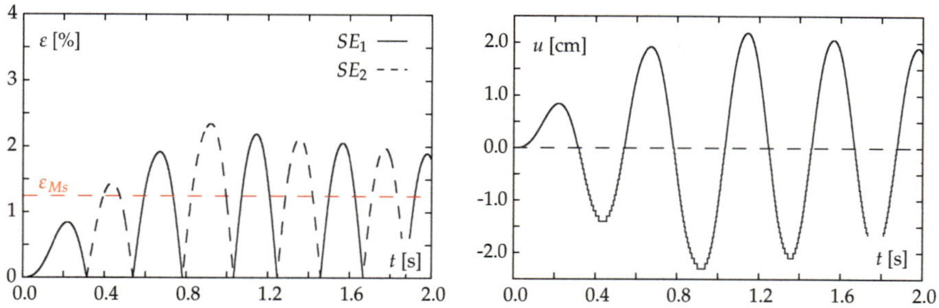

(a) Strain time history (b) Displacement time history

Fig. 12. Passive system with no pre-strain subjected to an harmonic load

can count with the contribution of both wires during the whole dynamic process. Moreover, one can set the initial strain to a level that can easily enable higher martensite transformation ratios and hence, higher damping (Zhang & Zu, 2007).

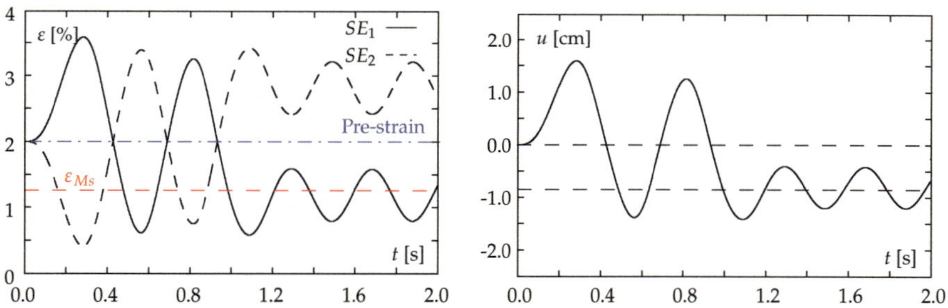

(a) Strain time history (b) Displacement time history

Fig. 13. Passive system with pre-strain subjected to an harmonic load

The effect of a 2.0% pre-strain in the wires of the previous passive system is illustrated in Figure 13, considering the same harmonic excitation. In this case, both of the SE wires are always active and they develop further martensitic transformation up to values of about 40 % martensite transformation ratio, meaning that the system is taking better advantage of the material's damping capabilities. However, two important drawbacks can be identified. First, this particular wire arrangement, with two pre-strained SE wires working in phase opposition, leads to a system with no recentering capability (Dolce et al., 2000). This can be confirmed in the displacements time history, during the steady state free vibration phase, when the oscillations are no longer centered in the original equilibrium position. The second, no less important drawback, relates to a variety of stress time effects that have been observed in various SMAs, that tend to adversely impact technological applications based on permanent pre-strained SE wires. According to the work presented by Auguet *et al.* (Auguet et al., 2008), in case of loading under stress controlled conditions, in the coexistence zone, stress seems to decrease under constant strain. The fact that pre-stress gradually fades away with time becomes a major limitation of this type of devices for seismic applications.

Another phenomenon that might hinder the use of pre-strained SE wires in passive dissipation devices is the possible occurrence of permanent deformation due to cumulative SMA creep (Torra et al., 2007). When this happens, the length of the SE wire increases continuously with cycling, meaning that, for a given amplitude of the oscillations, the strain is reduced, along with the amount of dissipated mechanical energy. Cumulative creep in NiTi is linked with the introduction of dislocations and other lattice defects generated at high stress, during the preceding loading cycle (Miyazaki et al., 1986; Moumni et al., 2005). Hence, this accumulated deformation may be controlled by keeping the SE wire inside the recoverable limits of a so called pseudo-elastic window (PEW). These limits, which ensure an appropriate material behavior, are determined by the material composition and its thermo-mechanical treatment. In fact, the stability of the deformation behavior may be improved by raising the critical slip stress. Thermomechanical heat treatments aimed to stabilize the superelastic behavior of NiTi have been found effective (Miyazaki et al., 1986), minimizing cumulative creep and avoiding the modification of the hysteretic cycle.

While a simple solution, involving a third elastic element, exists in order to recover the re-centering capability of the system (Dolce et al., 2000; 2005), the remaining problems related to stress time effects and cumulative creep do not have a simple solution in a passive device configuration. Stress time effects can not be avoided, as the use of permanent pre-strained SE wires is a must in order to obtain competitive damping ratios. In what concerns the cumulative creep, keeping the strains inside the PEW is a very challenging task when dealing with arbitrary seismic excitation.

4.3 Semi-active system

The proposed semi-active control device minimizes the SMA rheological effects by controlling the strain in the SE wires. The strain is self-adjusting, allowing the wires to become strain/stress free, when not subjected to a dynamic excitation. The system is also able to keep the wires deformation inside a given PEW, while guaranteeing a minimal threshold to their strain level. The strain in the SE wires is calibrated by controlling the displacements of the wires at their supports. The two wire supports can assume, independently, two configurations, locked or unlocked. By default, the supports are locked, assuring the adequate restraining for the SE wires. If the system needs to compensate for an excessively low or high strain in a given wire, it momentarily unlocks the wire through a controlled velocity action and without introducing additional forces into the system. As the actuating elements only have two fixed positions, this type of control system is usually referred to as a two-position or on-off controller.

Each of the two SE wires is independently controlled. The output signal from the controller associated with the SE wire i is defined as $U_i(s)$. Each controller has two reference inputs, $R_i^U(s)$ and $R_i^L(s)$ corresponding to the strain upper and lower bound limits, which, in this case, are constants. The controlled process is defined by the transfer function $G_{cp\,i}(s)$. The output $C_i(s)$ is fed back to the summing points, where it is compared with the reference inputs $R_i^U(s)$ and $R_i^L(s)$, yielding the actuating error signals $E_i^U(s) = R_i^U(s) - C_i(s)$ and $E_i^L(s) = R_i^L(s) - C_i(s)$. There is no need for a feedback-path transfer functions, to modify the output in order to make it comparable with the reference input signals, since the output of the numerical simulation directly yields the updated strain value for each SE wire. The block diagram of the on-off control system for one SE wire is presented in Figure 14(a). The signal $u_i(t)$ may

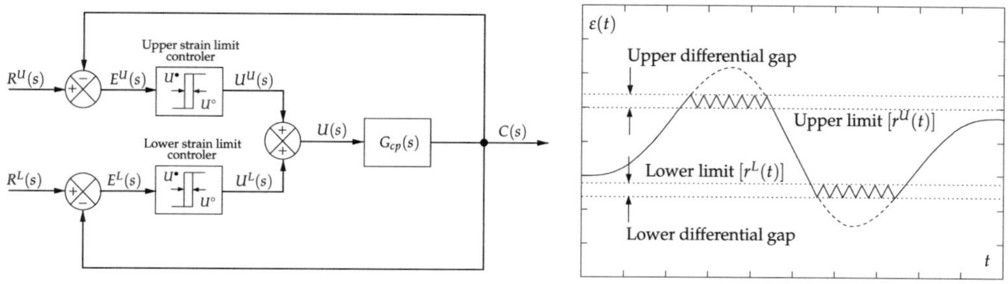

(a) Block diagram (b) Strain time history in a controlled wire

Fig. 14. On-off control system

assume the values $U_i^\bullet$ and U_i°, corresponding to the locked and unlocked configurations of the SE wire's supports, so that

$$\begin{aligned} u_i(t) = U_i^\bullet \text{ (locked)}, \quad &\text{for} \quad e_i^U(t) > 0 \quad \text{or} \quad e_i^L(t) < 0 \\ u_i(t) = U_i^\circ \text{ (unlocked)}, \quad &\text{for} \quad e_i^U(t) < 0 \quad \text{or} \quad e_i^L(t) > 0 \end{aligned} \tag{1}$$

The range through which the actuating error signal varies before the switching occurs is called a differential gap. By limiting the amplitude of the differential gap, one obtains a system with less oscillations. In this numerical simulation, the differential gap is related to the size of the implemented time step. The time history of the strain in a controlled SE wire is illustrated in Figure 14(b).

Analyzing the plots in Figure 15, one can see that the semi-active system is able to bound the strains in the SE wires between an upper and a lower strain limit, which, for this case study, are set to 3.0% and zero, respectively. When, during the dynamic oscillations, the strain in a given wire reaches one of these values, the corresponding support is unlocked until the mass displacement is reversed.

The process of unlocking and subsequent locking of a SE wires allows the system either to increase their cumulative strain, when compensating for excessively low strains, or to decrease them, when compensating for excessively high strains. As the compensation for lower strains is larger and usually more frequent than the compensation for higher strains, the final accumulated strain in the SE wires is positive. For the current case study the final accumulated strain in the system adds up to about 1.5%. Note that, there is no need for additional energy input to enforce the strain the SE wires, as the system uses the energy of the excitation itself. As the dynamic displacements develop, the system uses them to gradually increase the strain in the SE wires, inside the established PEW. When a wire is unlocked, the system's stiffness decreases momentarily, leading to slighter higher displacements amplitudes when compared with it's pre-strained counterpart. By limiting the strains, the system might also limit the martensite

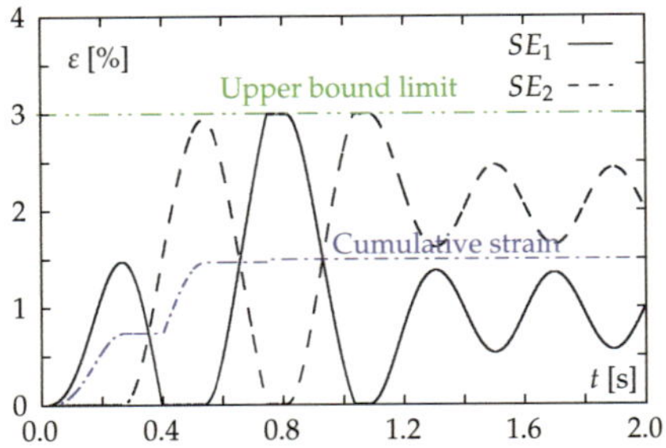

(a) Strain time history

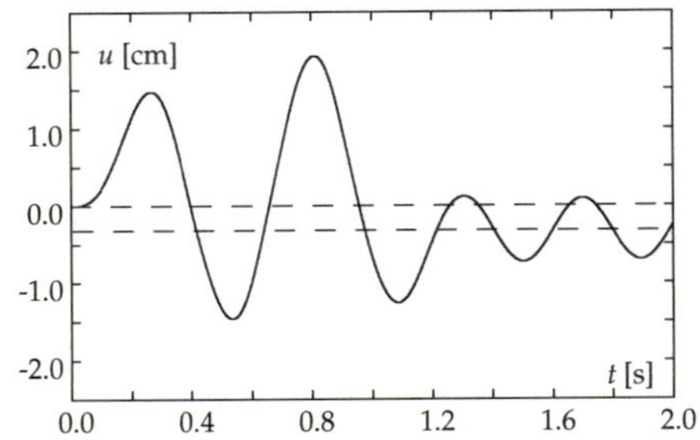

(b) Displacement time history

Fig. 15. Semi-active system subjected to an harmonic load

transformation ratio in the SE wires (in this test $\xi_{max} = 30\,\%$) and, hence, the amount of energy dissipated per time. When this happens, the system takes longer to attain the steady state free vibration phase. However, the oscillations amplitude in this last stage is of the same order of magnitude as in the pre-strained system, see Figures 13(b) and 15(b).

To emphasize its reliability, the semi-active system is excited next with a similar harmonic ground acceleration, of doubled amplitude, 2.5 ms^{-2}. To obtain a clearer picture of the steady state forced vibration phase, the duration of the excitation was also incremented from 1 to 3 seconds. The results of the numerical simulation presented in Figure 16 show that, although the displacements exhibited by the system increase accordingly, the strains in the SE wires continue to be bounded by the defined strain limits. The stress-strain hysteretic cycles presented in Figures 16(c) and 16(d) for the two SE wires show that, as a natural consequence of the applied strain constraints, the stresses in the corresponding SE wires are bounded as well. It is known that the stress levels, related to the beginning and ending of the forward and inverse transformations, are temperature dependent. For this particular loading frequency, during the initial mechanical loading cycles, the mean temperature of the SE wires decreases, until it stabilizes around a certain value. This phenomenon causes, during this transitional phase, a decrease of the stress upper bound limit for both SE wires. This phenomena can be observed in Figure 16(e), side by side with the system's force-displacement response curve, Figure 16(f).

As already mentioned, the original passive system that served as starting point for the present proposal of a semi-active system, has no recentering capability, a mandatory feature for seismic applications. To accomplish this requirement, a third restoring element is usually introduced (Dolce et al., 2000; 2005). This element enhances the recentering capability of the system, driving it back towards it's original equilibrium position. Assuming it has a pure elastic behavior, the force it delivers is proportional to the final displacement of the system. It is the balance between this force and the total stiffness of the two sets of SE wires when the system drifts away from it's final equilibrium position, that defines the final equilibrium position of the system. This element can have a simple elastic behavior or it can also be a SE wire, further increasing the damping capabilities of the system. In the proposed semi-active control device the recentering capabilities can, under certain conditions, be assured by it's ability to control the displacements of the wires at their supports. To emphasize this aspect, one considers again

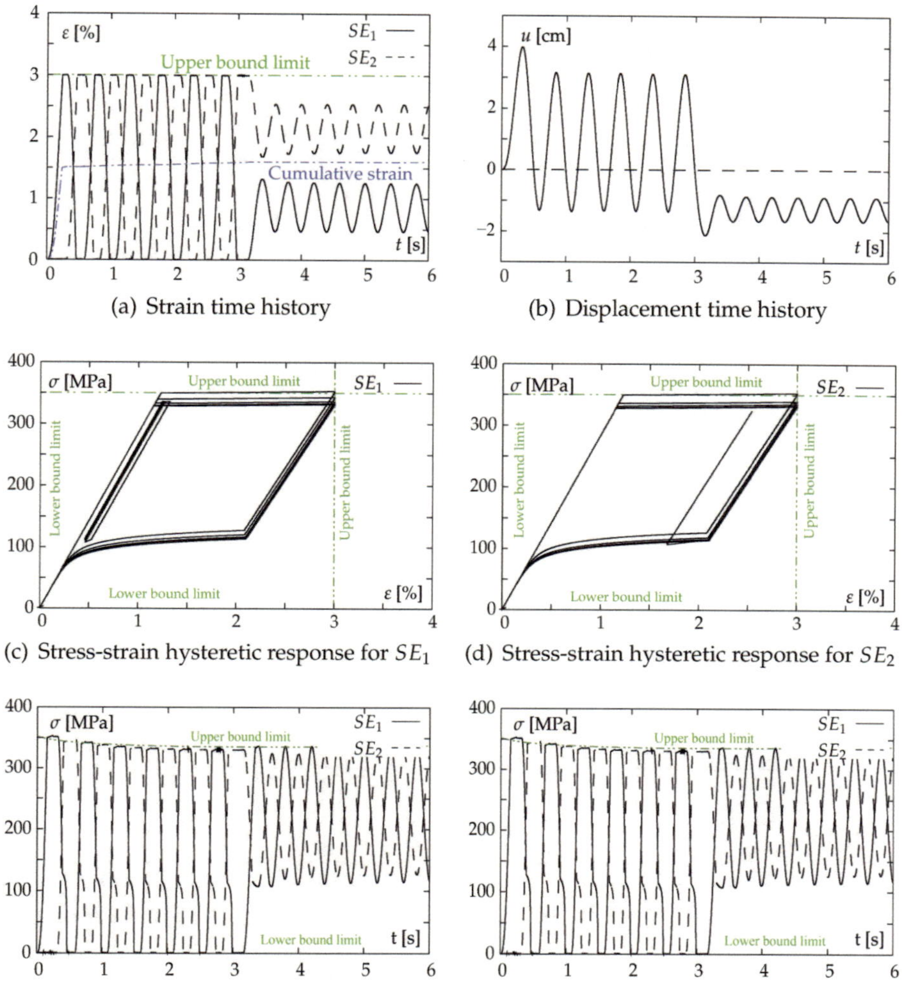

(a) Strain time history
(b) Displacement time history
(c) Stress-strain hysteretic response for SE_1
(d) Stress-strain hysteretic response for SE_2
(e) Stress time history
(f) Force-displacement hysteretic response

Fig. 16. Semi-active system subjected to an harmonic load of higher amplitude

the 1 second, 2 Hz harmonic ground acceleration with an amplitude of 1.25 ms^{-2} and focus on the response of the proposed device, once it reaches its steady state free vibration phase. Analyzing the displacement time history in Figure 17(b), one can see that, at the end of the loading period, the system oscillates around a new equilibrium position. To recover the initial equilibrium position, the strains/stresses in the SE wires are gradually set free, by unlocking the wires supports. During this process, the hysteretic loop is closed and the energy accumulated during straining is set free. This gives the system an extra damping capability that enables it to further decrease the amplitude of the free oscillations.

However, if the displacements amplitude exceeds a certain, pre-defined level during the loading period, the releasing of the strain/stress in the SE wires might not imply recovering of

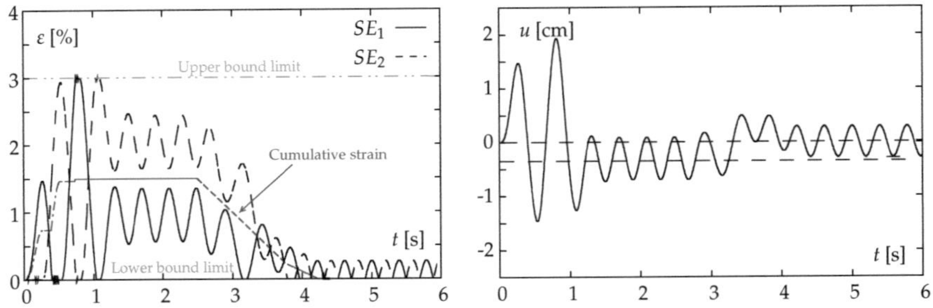

(a) Strain time history (b) Displacement time history

Fig. 17. recentering of the semi-active system subjected to an harmonic load

the initial equilibrium position. To regain the system recentering capability, a third elastic element should be used. To illustrate this situation, the relative displacements with respect to their initial positions, of two markers, M_1 and M_2, situated on the SE wires adjacent to the corresponding supports, see Figure 11(b), are plotted in Figure 18 for three different loading cases. The first graph, Figure 18(a), corresponds to the loading case that produces the dynamic

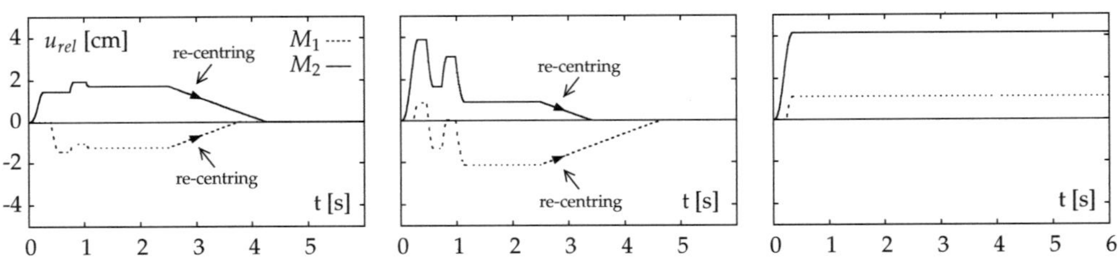

(a) First loading case (b) Second loading case (c) Third loading case

Fig. 18. Relative displacements of the SE wires at the supports.

response presented in Figure 17. One can see that, throughout most of the test, M_1 and M_2 move downwards and upwards, respectively, as the wires are released to compensate for low strains. When high strains are needed to be compensated, M_1 and M_2 reverse their movement but remain, however positive and negative, respectively. During the steady state free vibration phase, if the system releases the wires until M_1 and M_2 reach their initial position, the system is able to re-position itself in the original position, as already shown in Figure 17(b). In the second loading case, when the amplitude of the dynamic loading is doubled, while compensating for high strains, situations occur where M_1 becomes positive, as illustrated in Figure 18(b). However, as the system reaches its steady state free vibration phase with M_1 negative and M_2 positive, is still able to re-position itself. This is no longer the case if, with a different duration of the load, the system reaches its steady state free vibration phase in a situation where both M_1 and M_2 are positive, see Figure 18(c). From this configuration it is impossible to recover the original system position by just releasing the wires. This action will only cause further displacements in the positive direction of marker M_1. Therefore, one can conclude that, for similar situations, when the amplitude of the oscillations exceeds a predefined value, the system might loose its recentering capabilities. If one want to guarantee the recentering capabilities for virtually any loading, a third elastic element should be used in system, as in the case of passive control devices (Dolce et al., 2000; 2005).

4.3.1 Functioning of the proposed semi-active system

To better explain the sequence of actions taken by the proposed semi-active system during a dynamic excitation, one consider a typical displacement time-history response to an harmonic excitation, see Figure 19.

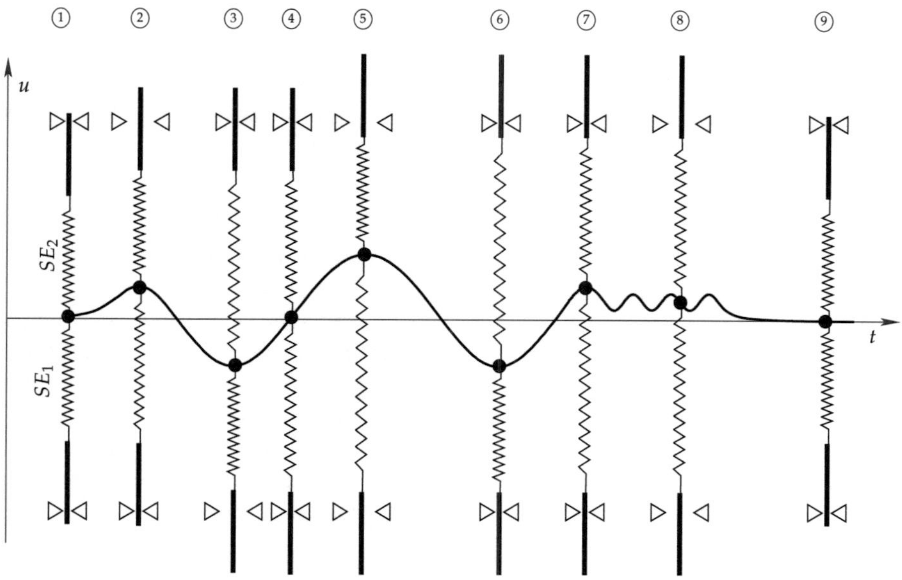

Fig. 19. Functioning of the proposed semi-active control system

The analysis starts with the system resting in it's original equilibrium position, referred as position ① in the graph. In this position both of the SE wires, SE_1 and SE_2, are strain/stress free and their supports, S_1 and S_2, locked. Once the harmonic excitation is applied to the system, the mass starts to move (assume) upwards and the straining of SE_1 begins. During the first half cycle of the movement, between position ① and ②, S_2 is unlocked to prevent compression in the corresponding wire, while the support of SE_1 is kept locked, to ensure the straining of the wire. As the mass reverses its direction, S_2 switch to locked, to preserve the accumulated strain in the wire. As the movement continues downwards, the support of SE_1 remains locked as well, until the mass reaches it's original equilibrium position. With this configuration, displacements below this point would cause compression in the SE_1. For this reason, as soon as this happens, S_1 switches to unlocked, until position ③, when the mass reverses once again its direction. At the end of the first loading cycle, position ④, both SE wires have already accumulated a certain level of strain. Note that this strain was introduced in the system by the external dynamic load itself, without any need of supplemental energy input. Between positions ④ and ⑤, assuming that in the last one occurs the maximum displacement amplitude, both supports will switch from locked to unlocked: S_2, as the strain in SE_2 approaches zero to avoid compression, and S_1, as the strain in SE_1 goes beyond the defined value inside the PEW, preventing it from further straining. If, during an eventual steady state forced vibration phase, positions ⑤ to ⑥, there is no need for adjustments in the SE wires to compensate low or high strains, both supports remain locked. As soon as the dynamic excitation ceases, position ⑦, the amplitude of the free vibrations starts gradually to diminish because of the SE damping. The system reaches it's steady state free vibration phase, position ⑧, oscillating around an equilibrium position different from it's original one. During this period, both supports remain locked. At this point, to recover the original equilibrium position, the system unlocks both of the cables, relinquishing the accumulated strain in the wires. Finally, position ⑨, the supports are locked again to prepare the device for the next dynamic excitation.

Note that, every time the system needs to unlock a support, the wire is released with a controlled velocity, which can be seen as a design parameter of the proposed semi-active device. In all the harmonic tests reported here, this velocity was set to 0.6 m/s.

5. Proposed semi-active vibration control device under seismic excitation

Studies on the efficiency of SMA based passive control devices in the context of seismic loading were reported in (Cismasiu & Santos, 2008), considering a simplified numerical model of the São Martinho railway viaduct, see Figure 20. This pre-stressed concrete structure, with a total length, between abutments, of 852.0 m, is built up of seven, 113.6 m length, independent segments and one segment of 56.8 m, adjacent to the south abutment. These segments are divided into 28.4 m spans and are structurally independent. The railway deck is a 13.0 m wide beam slab, comprising two 2.0×1.4 m main girders. The foundations are materialized by 1.2 m diameter piles with an average length of about 30.0 m. The concrete piers are tubular and have an average height of 12.0 m. Each pier is supported by five piles.

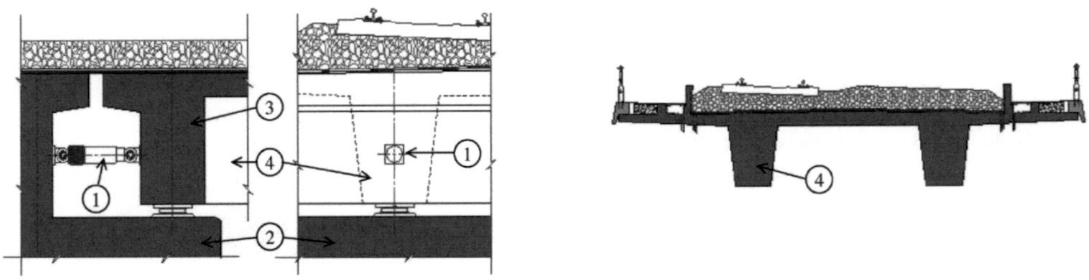

Legend: 1. Control device, 2. Abutment, 3. Transverse girder, 4. Main girder

Fig. 20. São Martinho railway viaduct: Control device location and mid-span cross section

For the longitudinal analysis of the viaduct, it is assimilated to an elastic, one degree-of-freedom dynamic system, with 4650 ton mass and a stiffness of 355×10^3 kN/m, having a structural damping of 5 %. The records of two historic strong earthquakes, *El Centro*, with a magnitude of 7 M_W and *Kobe*, with a magnitude of 6.9 M_W, as provided by the Pacific Earthquake Engineering Research Center and the University of California in the PEER Strong Motion Database (Pac, http://peer.berkeley.edu/smcat), are used in the numerical simulation.

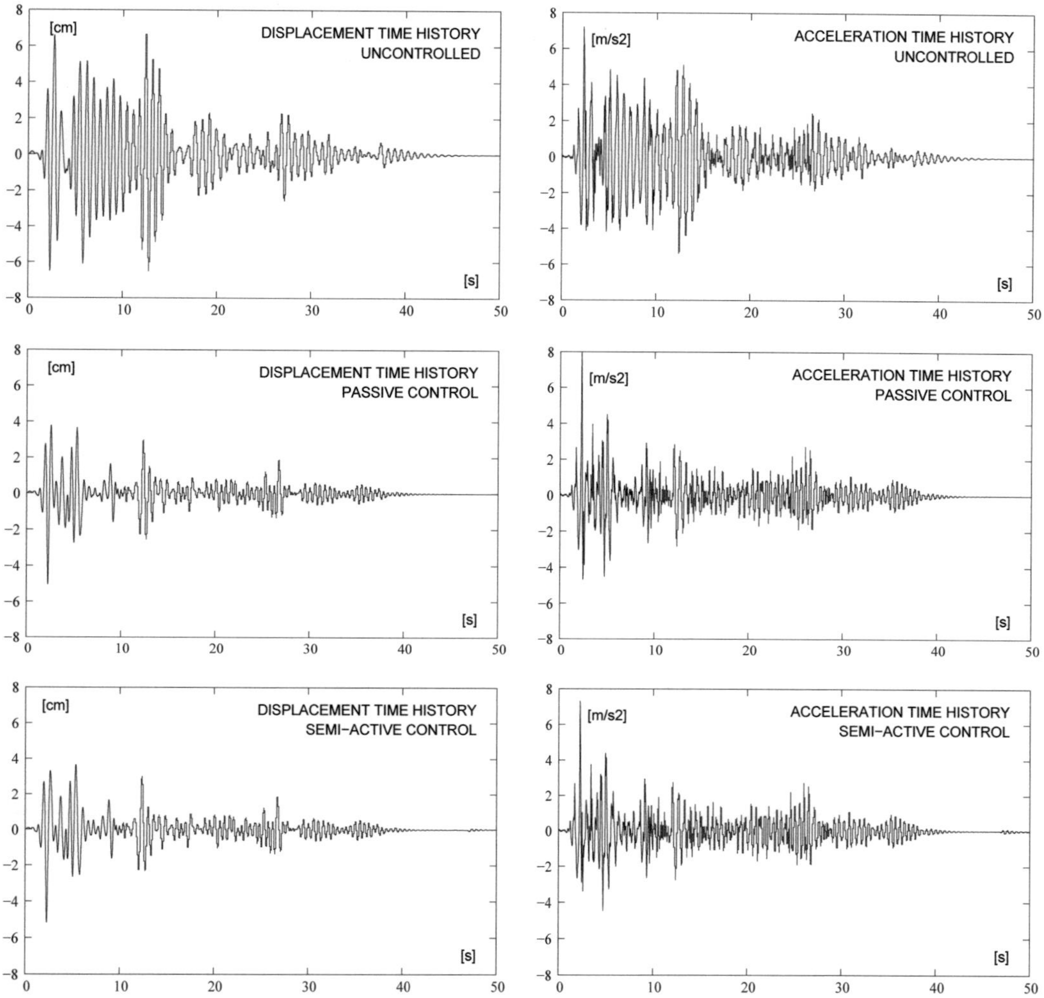

Fig. 21. Response of the structure to the *El Centro* earthquake: displacement and acceleration time history for the free and controlled structure.

To emphasize the benefits of the proposed semi-active control device on the structural response of the structure, three cases are considered: the uncontrolled structure, the structure controlled with a SMA based passive device and the structure controlled with the semi-active device. Note that, in order to get comparable sets of results, the strain level in the SE wires of the passive device was set equal with the maximum strain level attained in the semi-active device. To obtain an adequate response of the semi-active device to the higher frequency content of the seismic action, when the supports are unlocked, the wires are released with a velocity of 1 m/s. Two control devices, both either passive either semi-active, are placed at one end of the viaduct, one for each main girder. Each of them is composed of two sets of 1.0 m SMA wires, each set with a total area of 100 cm^2. This section could be built up of bars or a set of smaller wires laid parallel in strands, to form a cable. The mechanical characteristics of the SMAs are the ones defined in Table 1. A pre-strain of 2.25 % is considered in the passive

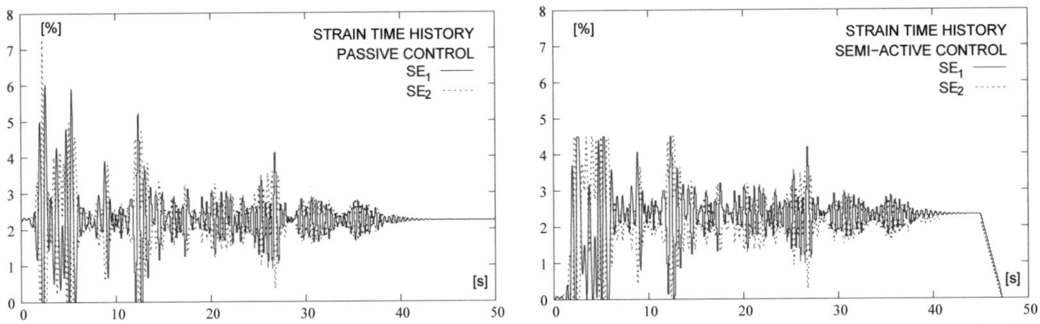

Fig. 22. Response of the structure to the *El Centro* earthquake: strain time history in the SE wires of the controlled structure.

device, equal to the resulting cumulative strain in the semi-active device. The displacement and acceleration time histories of the viaduct deck in the longitudinal direction, as well as the strain time history in the SE wires of the passive and semi-active control system, obtained when the structure was subjected to *El Centro* earthquake, are presented in Figures 21 and 22, respectively.

Analyzing the plots in Figure 21, one can see that, during the seismic excitation, the two control devices exhibit similar damping characteristics, being able to considerably reduce the amplitude of the earthquake induced displacements and accelerations. The plots in Figure 22 however, illustrate that the strain time history in the SE wires is quite different for the two adopted control solutions. Although the pre-strain in the passive device is equal to the cumulative strain in the semi-active device, one can see that the maximum strains developed in the passive device are considerably larger when compared with the ones developed in its semi-active counterpart. As their values are outside the PEW, the long term behavior of the passive device might be compromised. However, no failure occurs in the SE wires during the earthquake. One can observe that, when the seismic excitation stops, the SE wires return to their initial pre-strain, favoring the occurrence of relaxation phenomena. On the other hand, in the semi-active device, the wires return to their strain free condition at the end of the loading.

To conclude, one may say that, for the given seismic excitation, except for the long term relaxation phenomena, the two control solution produce equivalent results. One must note, however, that the pre-strain level in the passive control device was calibrated according to the given seismic excitation, in order to avoid failure in the SE wires. If a different seismic action is applied to the same structure configuration, the results might be completely different. In order to illustrate this behavior, the structure is subjected next to the *Kobe* earthquake.

Analyzing the dynamical response of the structure presented in Figures 23 and 24, one can see that the passive control device fails, as its SE wires' ultimate strain capacity ($\simeq 10$ - 12%) is exceeded. Besides relaxation and creep potential problems, the fact that this type of solutions need an *a priori* value for the level of pre-strain in the SE wires, represent an important drawback in the case of seismic applications. While estimates for the maximum strain in the SE wires are relatively easy to obtain in the case of service loads, and therefore a corresponding

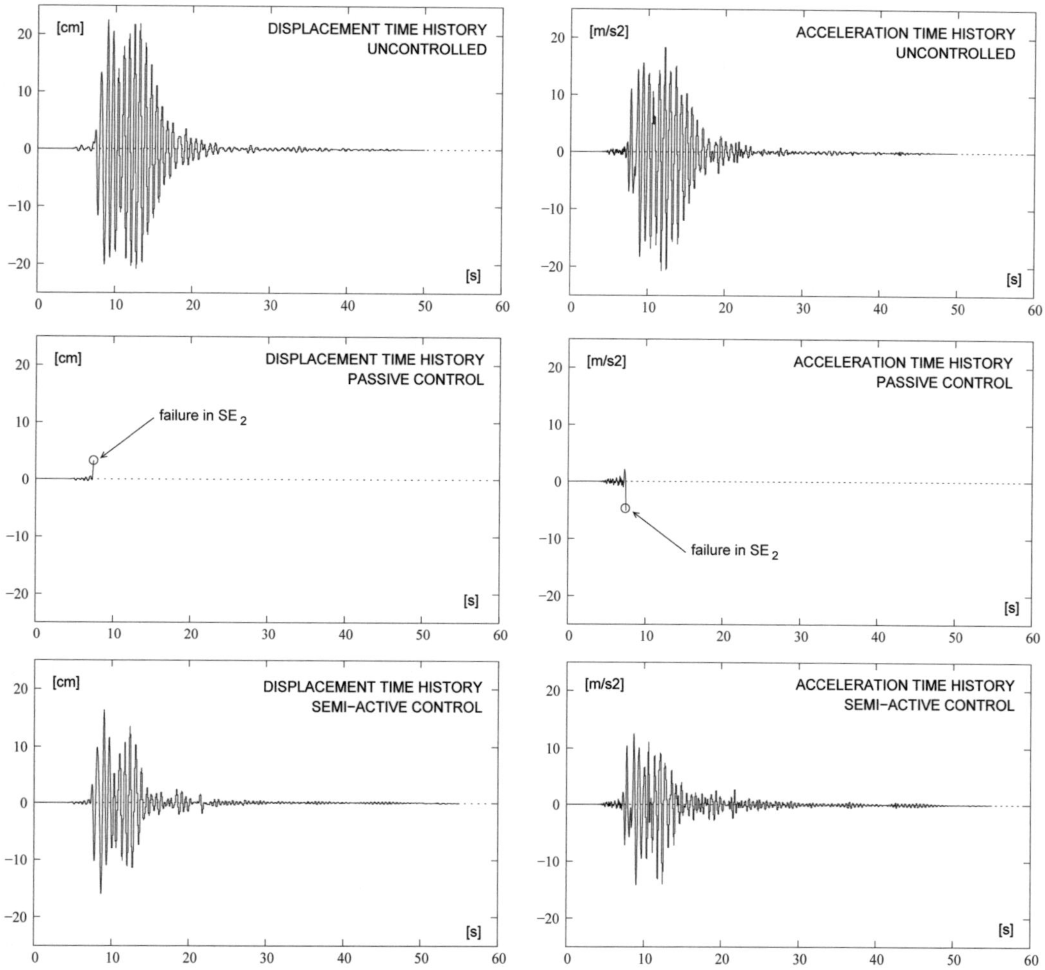

Fig. 23. Response of the structure to the *Kobe* earthquake: displacement and acceleration time history for the free and controlled structure.

value for the necessarily pre-strain can be computed, this is no longer the case for extraordinary dynamic loads.

With no need of initial pre-strain calibration, its semi-active counterpart responds well to virtually any level of dynamic excitation, as illustrated here in the case of the *Kobe* earthquake. When the response of the structure is compared with the one obtained for the *El Centro* earthquake, one can see that the system exhibits similar characteristics. It presents important damping capabilities, is able to confine the strains in the SE wires to predefined levels, inside the PEW, and finally, at the end of the action, is able to recover the SE wires strain free condition.

Another benefic aspect related to the semi-active control device is its ability to confine the force values throughout the entire duration of the seismic action, meaning that, when implemented in a civil engineering structure, the force the semi-active device transmits to the structure

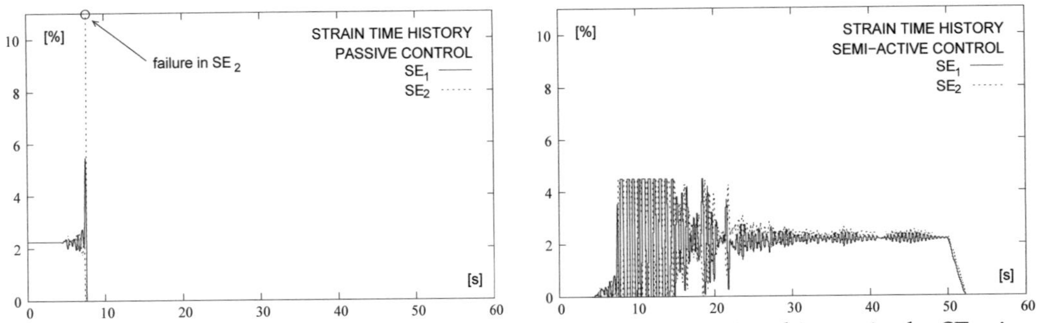

Fig. 24. Response of the structure to the *Kobe* earthquake: strain time history in the SE wires of the controlled structure.

can be conveniently bounded. This feature is very important in the designing process of the structure itself.

6. Conclusions

The proposed semi-active vibration control device originates from a passive control system, based on SE austenitic wires. In its semi-active version, the system monitors the feedback measurements and based on this information, continuously adjust the strain in the SE wires in order to improve its dynamical characteristics.

The strain accumulation in the wires is a result of the motion of the structure itself, with no need of external energy input in the system. To avoid relaxation phenomena, the SE wires are set strain free at the end of the loading period.

The system is able to dissipate a considerable amount of energy, while keeping the SE wires inside the recoverable limits defined by the PEW, to minimize the rheological effects related to cumulative creep. It also guarantees a minimal threshold to the strain level in the SE wires to avoid compression and exhibit efficient recentering capabilities.

The reported numerical simulations clearly demonstrate the potential the proposed semi-active control device has, in improving the seismic response of civil engineering structures. These results are expected to be confirmed with experimental tests in the near future, as a small scale prototype of the device is currently being developed.

7. Acknowledgment

This work is part of the research developed in UNIC, the Research Center of the Department of Civil Engineering, Faculty of Science and Technology of the Nova University of Lisbon and partially supported by contract SFRH/BD/37653/2007 with Fundação para a Ciência e Tecnologia. Cooperation with Professor F. M. Braz Fernandes from CENIMAT, the Center of Materials Research of the same institution is gratefully acknowledged.

8. References

Auguet, C., Isalgue, A., Torra, V., Lovey, F. & Pelegrina, J. (2008). Metastable effects on martensitic transformation in SMA. Aging problems in NiTi, *Journal of Thermal Analysis and Calorimetry* **92**(1): 63–71.

Auricchio, F., Fugazza, D. & DesRoches, R. (2006). Numerical and experimental evaluation of the damping properties of shape-memory alloys, *Journal of Engineering Materials and Technology* **128**(3): 312–319.

Auricchio, F. & Sacco, E. (1997). A one-dimensional model for superelastic shape-memory alloys with different elastic properties between austenite and martensite, *International Journal of Non-Linear Mechanics* **32**(6): 1101–1114.

Auricchio, F., Taylor, R. L. & Lubliner, J. (1997). Shape-memory alloys: macromodelling and numerical simulations of the superelastic behaviour, *Computer Methods in Applied Mechanics and Engineering* **146**: 281–312.

Bhattacharyya, A., Sweeney, L. & G., F. M. (2002). Experimental characterization of free convection during thermal phase transformations in shape memory alloy wires, *Smart Material and Structures* **11**: 411–422.

Brinson, L. C. & Huang, M. S. (1996). Simplifications and comparisons of shape memory alloy constitutive models, *Journal of Intelligent Material Systems and Structures* **7**: 108–114.

Chopra, A. K. (2001). *Dynamics of Structures: theory and applications to earthquake engineering*, 2nd edn, Prentice-Hall, Inc., Upper Saddle River, NJ 07458.

Cismasiu, C. & Santos, F. P. A. (2008). Numerical simulation of superelastic shape memory alloys subjected to dynamic loads, *Smart Materials and Structures* **17**(2): 25–36.

Collings, E. W. (1995). *Materials Properties Handbook: Titanium Alloys*, ASM International.

DesRoches, R. & Delemont, M. (2002). Seismic retrofit of simply supported bridges using shape memory alloys, *Engineering Structures* **24**: 325–332.

Dolce, M., Cardone, D. & Marnetto, R. (2000). Implementation and testing of passive control devices based on shape memory alloys, *Earthquake Engineering and Structural Dynamics* **29**: 945–968.

Dolce, M., Cardone, D., Ponzo, F. C. & Valente, C. (2005). Shaking table tests on reinforced concrete frames without and with passive control systems, *Earthquake Engineering and Structural Dynamics* **34**: 1687–1717.

Höhne, G., Hemminger, W. F. & Flammersheim, H.-J. (2003). *Differential Scanning Calorimetry*, 2nd edn, Springer.

Holman, J. (2009). *Heat Transfer*, Mcgraw-Hill Series in Mechanical Engineering, McGraw-Hill Science/Engineering/Math.

Housner, G. W., Bergman, L. A., Caughey, T. K., Chassiakos, A. G., Claus, R. O., Masri, S. F., Skelton, R. E., Soong, T. T., Spencer, B. F. & Yao, J. T. P. (1997). Structural control. Past, present, and future, *Journal of Engineering Mechanics* **123**(9): 897 – 971.

Isalgue, A., Torra, V., Yawny, A. & Lovey, F. C. (2008). Metastable effects on martensitic transformation in SMA. Part VI. The Clausius-Clapeyron relationship, *Journal of Thermal Analysis and Calorimetry* **91**(3): 991–998.

Ivshin, Y. & Pence, T. J. (1994). A thermomechanical model for a one variant shape memory material, *Journal of Intelligent Material Systems and Structures* **5**(4): 455–473.

Jones, D. R. H. & Ashby, M. F. (1998). *Engineering Materials: An Introduction to Microstructures, Processing and Design*, Vol. 2 of *International Series on Materials Science and Technology*, 2nd edn, Butterworth-Heinemann.

Koistinen, D. P. & Marburger, R. E. (1959). A general equation prescribing the extent of the austenite-martensite transformation in pure iron-carbon alloys and plain carbon steels, *Acta Metall* **7**: 59–60.

Liang, C. & Rogers, C. A. (1990). One-dimensional thermomechanical constitutive relations for shape memory materials, *Journal of Intelligent Systems and Structures* **1**: 207–234.

Magee, C. L. (1970). Nucleation of martensite, *Phase Transformations, ASM* pp. 115–156.

Matsui, R., Tobushi, H. & Ikawa, T. (2004). Transformation-induced creep and stress relaxation of TiNi shape memory alloy, *Proceedings of the Institution of Mechanical Engineers, Part L: Journal of Materials: Design and Applications* **218**(4): 343–353.

McCormick, J., DesRoches, R., Fugazza, D. & Auricchio, F. (2006). Seismic vibration control using superelastic shape memory alloys, *Journal of Engineering Materials and Technology* **128**(3): 294–301.

Miyazaki, S., Imai, T. & Otsuka, K. (1986). Effect of cyclic deformation on the pseudoelasticity characteristics of Ti-Ni alloys, *Metalurgical Transactions A* **17A**: 115.

Moumni, Z., Van Herpen, A. & Riberty, P. (2005). Fatigue analysis of shape memory alloys: energy approach, *Smart Materials and Structures* **14**: 287–292.

Ocel, J., DesRoches, R., Leon, R. T., Hess, W. G., Krumme, R., Hayes, J. R. & Sweeney, S. (2004). Steel beam-column connections using shape memory alloys, *Journal of Structural Engineering* **130**(5): 732–740.

Ortín, J. & Delaey, L. (2002). Hysteresis in shape-memory alloys, *International Journal of Non-Linear Mechanics* **37**: 1275–1281.

Otsuka, K. & Wayman, C. M. (eds) (1998). *Shape memory materials*, Cambridge University Press.

Pac (http://peer.berkeley.edu/smcat). *PEER Strong Motion Database.*

Shook, D. A., Roschke, P. N. & Ozbulut, O. E. (2008). Superelastic semi-active damping of a base-isolated structure, *Structural Control and Health Monitorig* **15**: 746–768.

Symans, M. D. & Constantinou, M. C. (1999). Semi-active control systems for seismic protection of structures: a state-of-the-art review, *Engineering Structures* **21**: 469–487.

Tanaka, K., Kobayashi, S. & Sato, Y. (1986). Thermomechanics of transformation pseudoelasticity and shape memory effect in alloys, *International Journal of Plasticity* **2**: 59–72.

Torra, V., Isalgue, A., Martorell, F., Terriault, P. & Lovey, F. (2007). Built in dampers for family homes via SMA: An ANSYS computation scheme based on mesoscopic and microscopic experimental analyses, *Engineering Structures* **29**: 1889–1902.

Vitiello, A., Giorleo, G. & Morace, R. E. (2005). Analysis of thermomechanical behaviour of Nitinol wires with high strain rates, *Smart Material and Structures* **14**: 215–221.

Zhang, Y. & Zu, S. (2007). A shape memory alloy-based reusable hysteretic damper for seismic hazard mitigation, *Smart Materials and Structures* **16**: 1603–1623.

8

Seismic Vibration Control of Structures Using Superelastic Shape Memory Alloys

Hongnan Li and Hui Qian
Dalian University of Technology, Zhengzhou University
China

1. Introduction

Structural vibration control has been approved effective design strategy for enhancing structural performance against earthquake and mitigating seismic response. Passive vibration control is a promising technology which has been widely used in earthquake engineering due to its simple configuration, low cost, easy maintenance and reliable run without power support. However, current technologies present some limitations, such as problems related to aging and durability, residual displacement, substitution after strong events, among others. Recently, the increasing research and development of smart materials and controlling devices open up a new area for seismic vibration control of structural engineering, providing a basic platform for the design and exploration of new generation high-performance structural systems.

Shape memory alloys (SMAs) are a class of novel smart materials that possess unique properties, including shape memory effect, superelasticity effect, extraordinary fatigue resistance, and high corrosion resistance and damping characteristics, which make them perfect candidates for applications in seismic resistant design and retrofit of structures.

A significant number of research studies have been conducted in an effort to use SMAs for applications in seismic resistant design and retrofit of structures in the past decade. The results of the previous works show there is promise for SMAs in earthquake engineering. However, significant research is still needed, particularly as SMAs strands are planted into the concrete structures. The damping property and self-recentering capacity of the SMAs-based intelligent RC structure need to be further studied. In addition, to design some new types of passive seismic devices which possess not only energy dissipation and recentering capabilities but also simple configuration for easy installation in the practical engineering is still actual issue currently. Moreover, at present, there is little available data on the optimization of their size and position of the SMAs dampers in a structure as the economic cost is considered.

This chapter presents some applications of shape memory alloys in seismic resistant design. *Main contents are included as follows:*

2. Vibration Control of Concrete Structures Reinforced With Superelastic Shape Memory Alloy

2.1 Self-rehabilitation property of intelligent concrete beams reinforced with superelastic SMA strands

Cracks can seriously affect the durability of reinforced concert structures. SMA has the properties of shape memory effect, super-elasticity, and good corrosion resistance, which can change the properties of the concrete structures (Choi et al, 2008; Maji et al, 1998; Song et al, 2006). In this section, we focus on the self-rehabilitation property of intelligent concrete structures reinforced with superelastic SMA strands.

2.1.1 Specimens

Two concrete beams were fabricated and their dimensions and reinforcement arrangements are shown in Fig. 1. The reinforcements of specimen RC are steel wires with a diameter of 2.2mm. The reinforcements of specimen A are super-elastic SMA strands, which are composed of seven single wires with a diameter of 0.75mm. The mechanical properties of SMA strands and steel wires can be found in the Ref. (Li et al, 2006).

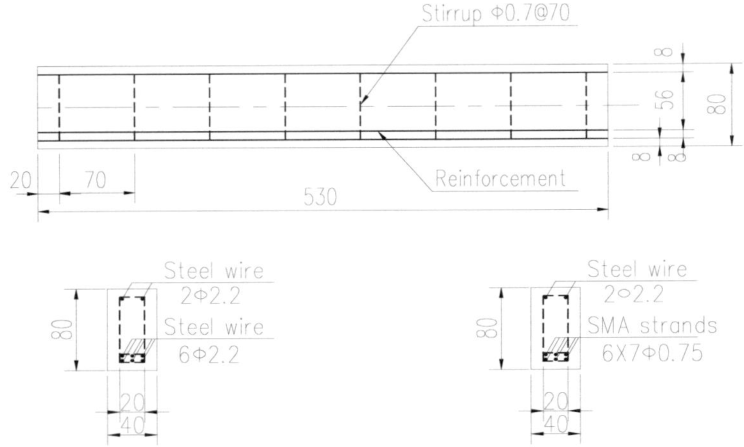

Specimen RC Specimen A

Fig. 1. Dimensions and reinforcement arrangements of specimens (Unit: mm)

2.1.2 Three-point bending test

In this experiment, the three-point bending tests were conducted on the specimens. The load is measured by a load cell in the test machine, the deflection at mid-span of the specimens is measured by a micrometer gauge, and the crack width is measured by an extensometer. The loading process is controlled by the displacement at mid-span and the loading rate is 0.2mm/min; the maximal displacement is set at 10mm.

2.1.3 Experimental results

The load-deflection curves of specimens are shown in Fig. 2. Specimen RC (Fig. 2(a)) shows obvious yield behavior after the 2nd loading cycle. The load-deflection curve of specimen A (Fig. 2(b)) shows hysteresis characteristics. After cracks are generated, the deformation of the SMA strands is very small, and the stress also small because its Young's modulus is just 1/5 of that of steel, and thus the bearing capability decreases immediately. With the increasing of the loading cycle numbers, the bearing capability of specimen A increase greatly; its bearing capability can reach to 6617.40N at the last cycle.

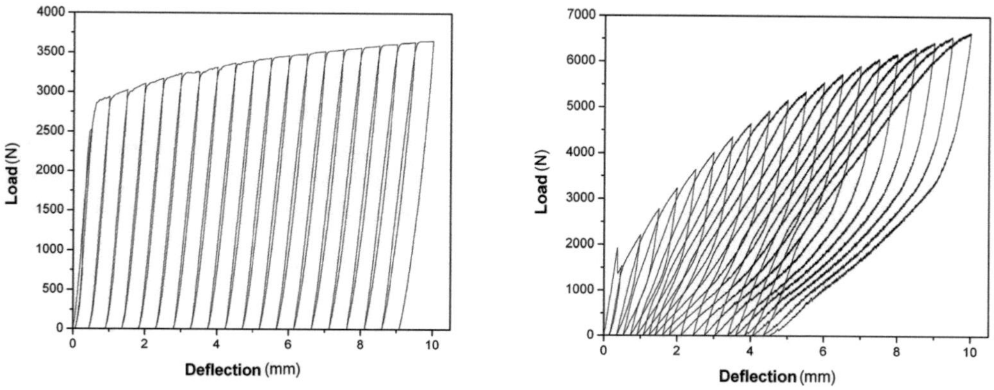

(a) Specimen RC (b) Specimen A

Fig. 2. Load-deflection curves of specimens

To illustrate the effects of super-elastic SMA strands in controlling the crack width, Fig. 3 shows the relationship between the residual crack width and the No. of loading cycles. The residual crack width of specimen RC and specimen A both increase with the increase of the loading cycles. Due to the superelasticity and the cementing bond between SMA strands and concrete, the crack width of specimen A is smaller than that of specimen RC.

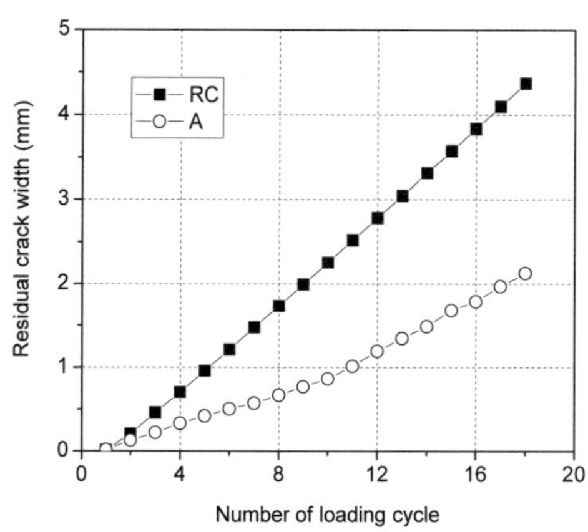

Fig. 3. Relationship between residual crack and number of loading cycles

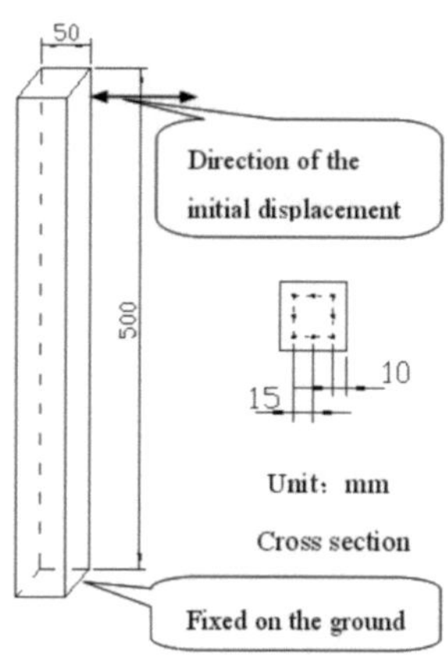

Fig. 4. Sketch of concrete column and details of its cross section

2.2 The influence of superelastic SMAs on the damping ratio of concrete columns

In the past several years, studies have shown the feasibility of using shape memory alloys for passive and active damping applications. The SMA energy dissipation devices have been proposed for bridges and buildings (Dolce et al, 2000; Wilde et al, 2000). Although studies have shown the feasibility of using SMAs as external energy dissipation devices, little has been done in using SMAs as inner energy dissipation devices to change the properties of concrete structures or structural members under seismic loadings or wind loadings in civil engineering. In this section, the reinforcements of the concrete columns are replaced by superelastic SMA wires to investigate the influence of SMAs as internal energy dissipation devices on the damping ratio of concrete columns.

2.2.1 Specimens

For comparison, the three kinds of concrete columns that were designed are as shown in Fig. 4. One is reinforced with eight steel wires (specimen S8), the other is reinforced with eight superelastic SMA wires (specimen A8), the third one is reinforced with eight martensitic SMA wires (specimen M), respectively. The mechanical properties of the steel wires and the SMA wires can be found in the Ref. (Cui et al, 2009).

2.2.2 Test instruments and method

To obtain the damping ratio of the specimens, the initial displacement method is utilized to measure the vibration curves of each specimen. The direction of the initial displacement is shown in Fig. 4. During the test, AR-5F acceleration transducer is used to measure the acceleration response of the top of the columns. All the test data are collected by a real-time data acquisition system named DSPS. For specimen M, martensitic SMA wires were strained up to 4% and connected with each other in series. During the test, the martensitic SMA wires were heated with a large current of 40A.

2.2.3 Test results

The normalized free vibration curves of the columns are shown in Fig. 5. The damping ratios of the specimens are listed in Table 1. From Fig. 5, it can be seen that after being electrified, the vibration curve of specimen M attenuates very fast, its damping ratio is almost the same with that of specimen S. The results demonstrate that the initial phase of SMA wires has great influence on the damping ratio of the structural members. Thus, the superelastic SMAs can be taken as an inter energy dissipation device to control the vibration of the structures.

2.3 Seismic response mitigation and the self-recentering capability of concrete frame with SMA strands

To investigate the capability of the self-damping and self-recentering of superelastic SMA reinforced concrete structure (SMA-structure), a two-layer eccentric structure was designed in this section as shown in Fig. 6, where, CK is the center of stiffness, and CM is the center of mass. The eccentricity is e_x=0.647m and e_y=0.0m. For reinforced concrete structure without SMAs (RC-structure), the rebar in the beams and columns are all steel; for the

SMA-structure, the rebar in beams are still steel, while the rebar in columns are superelastic SMA wires.

The seismic responses of structures are analyzed by using the finite element analysis program ANSYS. The SOLID185 of ANSYS is used to model the beams and columns of the frame structure. The MASS21 of ANSYS is used to model the weight of the structure.

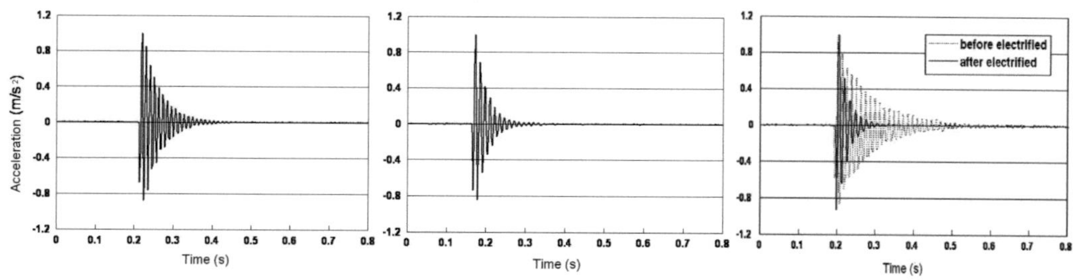

(a) Specimen S (b) Specimen A8 (c) Specimen M

Fig. 5. Free vibration curves of specimens

Table 1. Damping Ratios of Specimens

Specimen	S	A8	M	
			(before electrified)	(after electrified)
Damping ratio (%)	4.18	6.22	2.08	6.30

The seismic responses of the superelastic SMA reinforcedstructure and the RC-structure are analyzed in the time domain. The Ninghe, El Centro and Qian'an earthquake records are selected to be input to the structure as excitations, representing different site conditions. The peak accelerations of earthquake records are scaled to 0.4g ($g=9.8m/s^2$). The structural response of SMA-structure is expressed by R_{SMA}. Also, the response of RC-structure is expressed by R_{BC}. The response reduction ratio of the structure is defined as

$$\eta = \frac{R_{RC} - R_{SMA}}{R_{RC}} \times 100\% \tag{1}$$

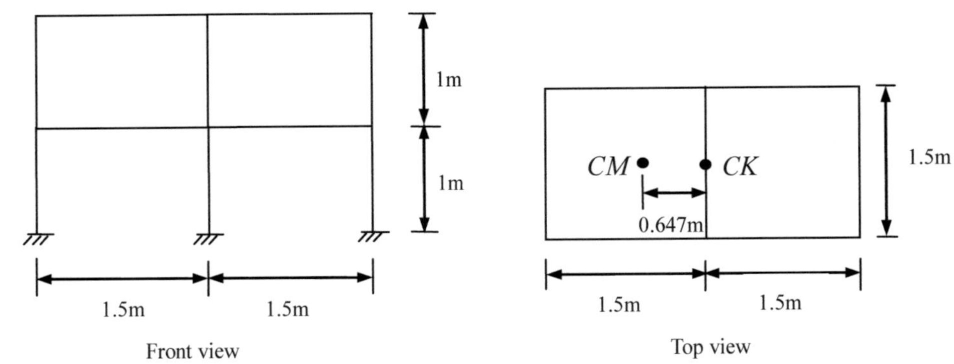

Fig. 6. Plan of the structure

The maximum displacements of the structures and the response reduction ratios are computed for three earthquake records. The results are listed from Table 2 to Table 4, where, η_{dx}, η_{dy} and η_{θ} are the reduction ratio in x, y and torsion direction. It can be seen from the tables that the responses of SMA-structure in each degree of freedom are almost less than those of RC-structure. However, the reduction ratios are different for the different earthquake records. The displacement time history curves of top story are shown from Fig. 7 to Fig. 9 for El Centro earthquake. It can be seen from these figures that the structural response are reduced in the whole time history under the influence of the SMA wires.

In order to investigate the self-restoring capability of SMA-structure, the position variation situation of the structure center under El Centro earthquake is shown in Fig. 10. It is obvious that superelastic SMA reinforced structure always move around the initial position. This phenomenon fully demonstrates the excellent self-restoring capability of SMA-structures compared to RC-structures.

Table 2. Maximum displacements of the structures (Qian'an)

	X (mm)			Y (mm)			θ ($\times 10^{-3}$ rad)		
	SMA	RC	η_{dx} (%)	SMA	RC	η_{dy} (%)	SMA	RC	η_{θ} (%)
1	2.9	3.2	16.75	3.1	3.8	19.07	2.36	2.89	18.37
2	5.0	4.8	-3.16	5.9	5.9	0.79	4.47	4.41	-1.25

Table 3. Maximum displacements of the structures (El Centro)

	X (mm)			Y (mm)			θ ($\times 10^{-3}$ rad)		
	SMA	RC	η_{dx} (%)	SMA	RC	η_{dy} (%)	SMA	RC	η_{θ} (%)
1	4.7	8.0	41.49	5.8	10.0	42.28	4.31	7.46	42.20
2	7.7	9.1	15.93	9.6	11.5	16.13	7.17	8.57	16.35

Table 4. Maximum displacements of the structures (Ninghe)

	X (mm)			Y (mm)			θ (×10-3 rad)		
	SMA	RC	η_{dx} (%)	SMA	RC	η_{dy} (%)	SMA	RC	η_{θ} (%)
1	4.3	29.9	85.74	5.3	37.2	85.74	3.97	27.99	85.81
2	6.2	31.1	80.16	8.0	39.5	79.76	5.90	29.53	80.02

3. Innovative Superelastic SMAs-based Re-centering Damping Device for Structural Vibration Control

3.1 Innovative Superelastic SMAs-based Re-centering Damping Device

In this study, innovative hybrid SMA friction device (HSMAFD) (Li et al, 2008) as energy dissipation device was considered. HSMAFD consists of recentering device based on pre-tensioned superelastic NiTi wires and friction device (FD), as shown in Fig. 11. The superelastic NiTi wires mainly perform a function of recentering due to large restoring

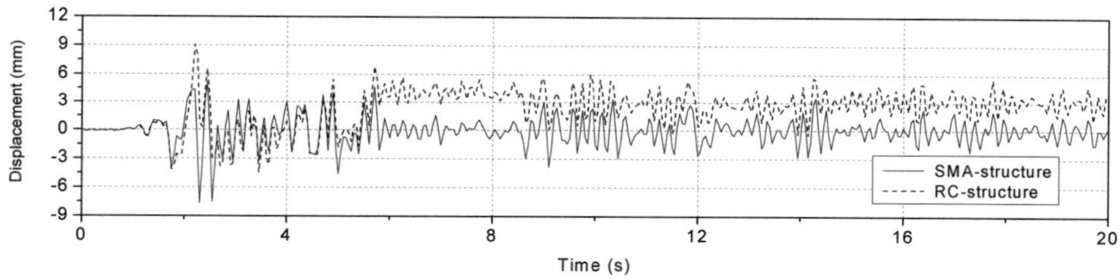

Fig. 7. Time history of the displacement in x direction of top story (El Centro)

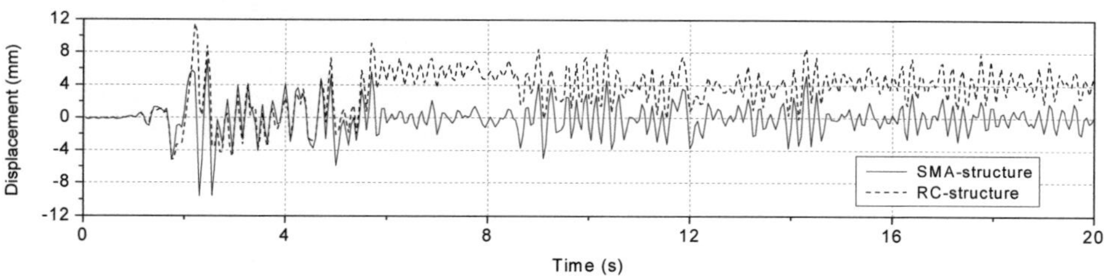

Fig. 8. Time history of the displacement in y direction of top story (El Centro)

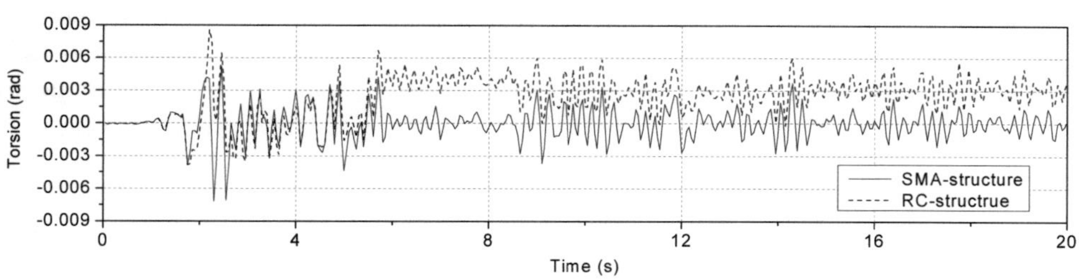

Fig. 9. Time history of the torsional displacement of top story (El Centro)

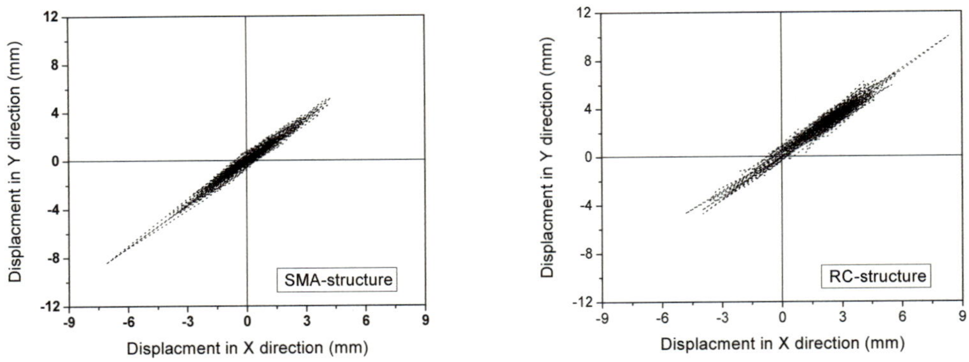

Fig. 10. Position change of structures under El Centro earthquake (0.4g)

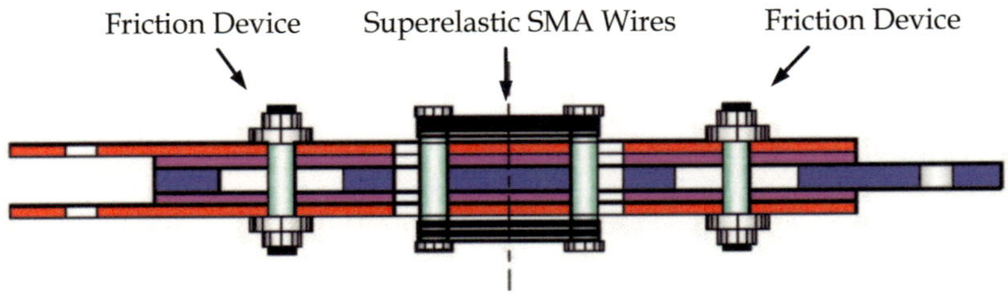

Fig. 11. Scheme diagram of a HSMAFD

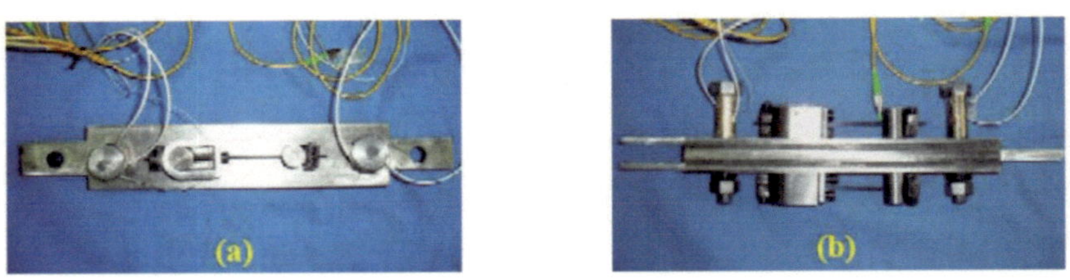

Fig. 12. Photos of HSMAFD

force, as well as additional energy dissipation owing to their hysteretic damping property. The friction devices are utilized to dissipate most seismic energy by relative sliding between the friction plants impacted by high strength bolts. By properly combining the number and the pre-tensioning level of the superelastic SMA wires and the friction force, the device exhibits supplemental recentering capacity and energy dissipating feature, which can be exploited to reduce structural seismic response and return the building to the initial position after earthquakes. Fig. 12 presents a HSMAFD model fabricated based on the design.

Fig. 13 shows hysteretic behavior of a HSMAFD under sinusoidal cyclic displacements of increasing amplitudes with loading frequency of 0.05Hz, 0.1Hz, 1.0Hz and 2.0Hz, respectively. As we can see, satisfying hysteretic loops, including both recentring and the energy dissipating features, were obtained under various conditions. With the increase of the displacement amplitude, the energy dissipation per cycle increases almost linearly, while the secant stiffness and the equivalent damping decrease. Moreover, the performance is not highly sensitive to frequencies of loading in the range of which earthquake engineering concerns.

3.2 Shaking Table Tests

3.2.1 Shaking table

The experimental tests were carried out on the earthquake simulator shaking table of the State Key Laboratory of Coastal and Offshore Engineering in Dalian University of Technology. It included a 3m×4m steel platform, with three degree of freedom, driven by servo-hydraulic actuators which were numerically controlled by a MTS analogue electronic control system. The shaking table has about 10 tons payload capacity with frequency band of 0.1~50Hz.

3.2.2 Structural model

The structural model in the tests is a 1/4 scaled three-story steel frame. Its overall dimension is 1.5×1.5m in plane and 2.4m in elevation. All the columns in the model have constant cross section (30mm×30mm×3mm angle iron). Similarly, all the beams have the same cross section (60mm×40mm×3mm rectangle iron). The main geometrical characteristics of the steel frame model are shown in Fig. 14. The yield stress and Young's modulus of the material are 215MPa and 206GPa, respectively. Masses added to the structure are 300.6kg for the first floor, 300kg for the second floor and 318.5kg for the third floor.

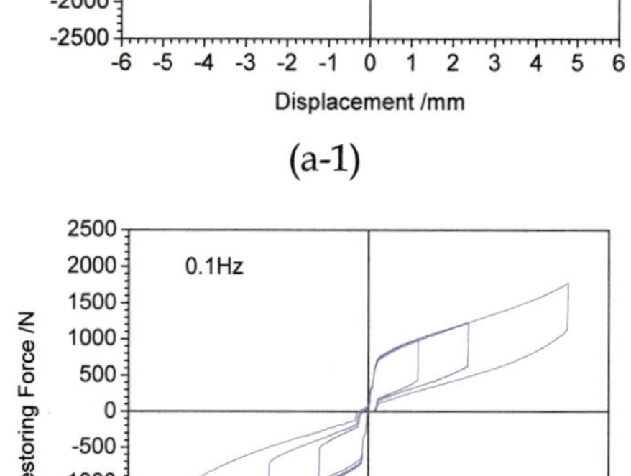

(a-1)

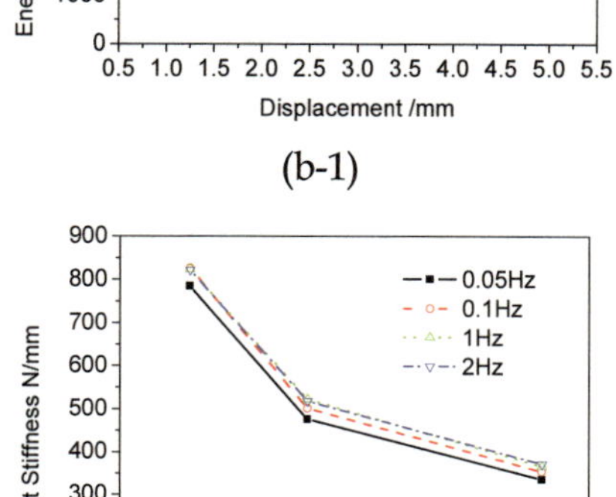

(b-1)

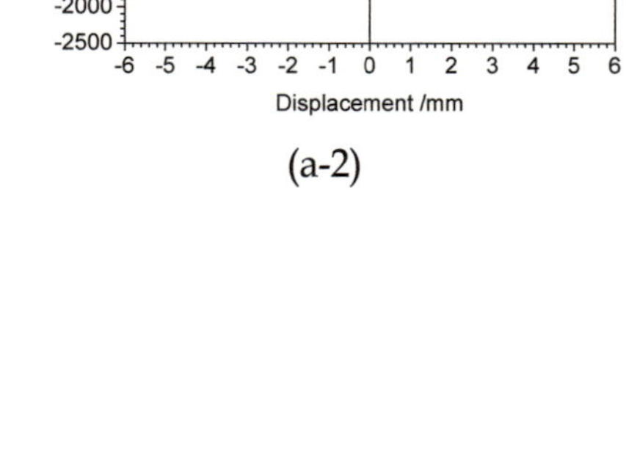

(a-2)

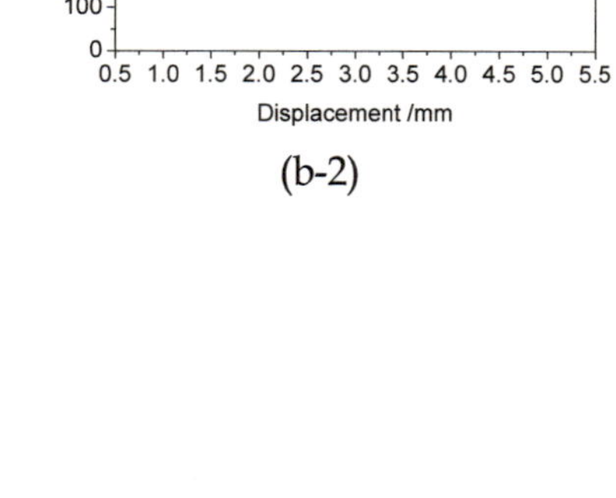

(b-2)

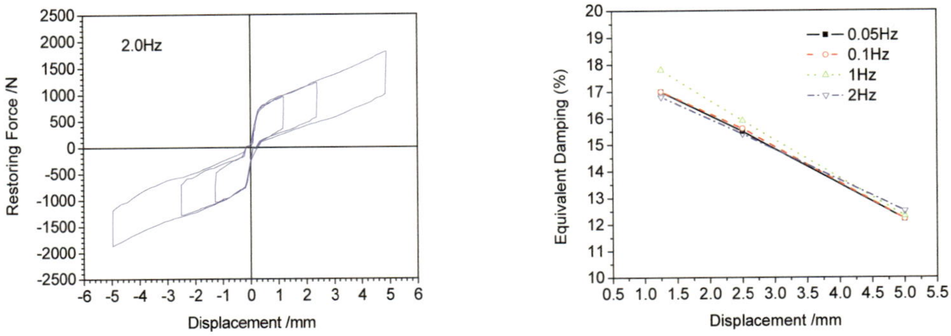

(a-3) (b-3)

Fig. 13. Hysteretic behaviour of HSMAFD under various loading frequency and displacement amplitudes: (a) Force-displacement curves; and (b) Mechanical quantities.

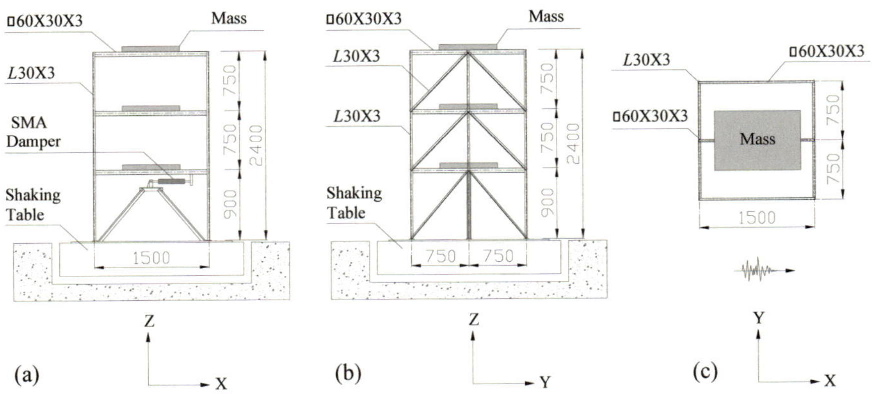

Fig. 14. Geometrical characteristics of the steel frame model (unit: mm):
(a) X direction elevation; (b) Y direction elevation; and (c) typical floor plan.

3.2.3 Earthquake historical records

In the tests, the flowing three earthquake historical records were selected: (i) El Centro. The N-S component recorded at the Imperial Valley Irrigation District substation in El Centro, California, during the Imperial Valley, California earthquake of May 18, 1940; (ii) Taft. The S69E component recorded at Taft Lincoln School Tunnel, Kern County, California earthquake of August 31, 1976; (iii) Tianjin. The N-S component recorded at the Tianjin hospital, Ninghe, Tangshan aftershock of November 15, 1976. The absolute peak acceleration of the earthquake records are adjusted to 0.15g. Additionally, the equally-spaced intervals of El Centro, Taft and Tianjin are 0.01s, 0.01s and 0.005s, respectively. Additionally, random white noise tests were conducted to identify the fundamental frequency of vibration and assess the damage suffered by the structure subjected to the seismic records.

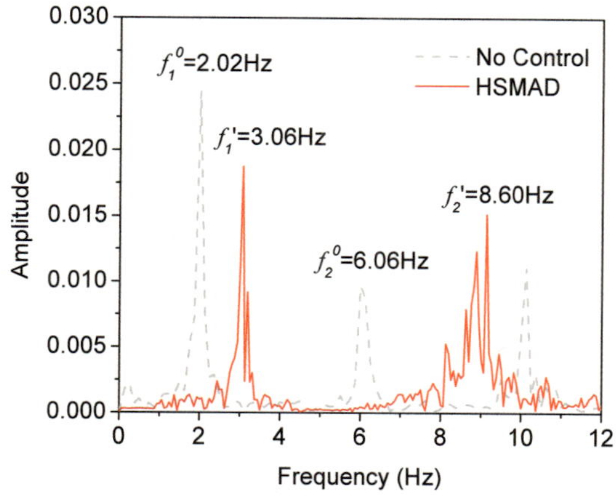

Fig. 15. Frequency spectrum of the free and controlled structures

3.2.4 Shaking table tests results

Fig. 15 shows the frequency spectrum of the structure with and without damping device. As can be seen, the fundamental frequency of the structure increases from 2.02Hz to 3.06Hz, up more than 50%, after HSMAFD was installed, which indicates that the energy dissipation system provides supplemental stiffness into the original structure.

Fig. 16 shows the profiles of maximum interstorey drifts for the free and controlled structures under different earthquakes with 0.15g PGA. As seen from the figure, interstorey drifts of the building have been non-identically reduced. Especially, compared with the two upper floors, the first floor has a more significant effect, decreasing by 69.1% for El Centro, by 65.5% for Taft and by 77.1% for Tianjin, respectively. This is because the damping device was only installed in the first floor of the building.

Fig. 17 shows the profiles of maximum absolute displacements relative to the table for the free and controlled structures under different earthquakes with 0.15g PGA. As can be seen, the maximum absolute displacements relative to the table of the top floor of the building decrease by 61.0% for El Centro, 45.7% for Taft and 65.6% for Tianjin, respectively.

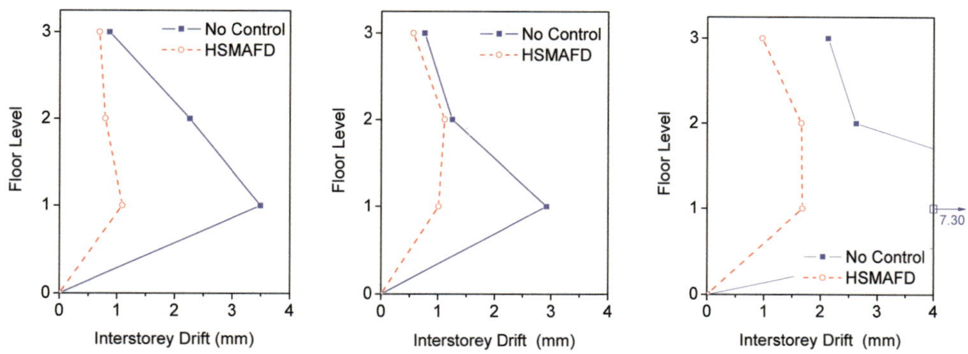

(a) El Centro (b) Taft (c) Tianjin

Fig. 16. Profiles of maximum interstorey drifts for the free and controlled structures

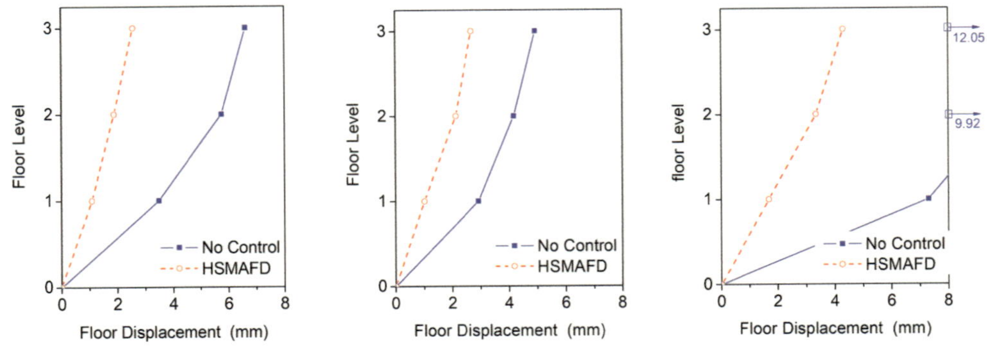

(a) El Centro (b) Taft (c) Tianjin

Fig. 17. Profiles of maximum absolute displacements for the free and controlled structures

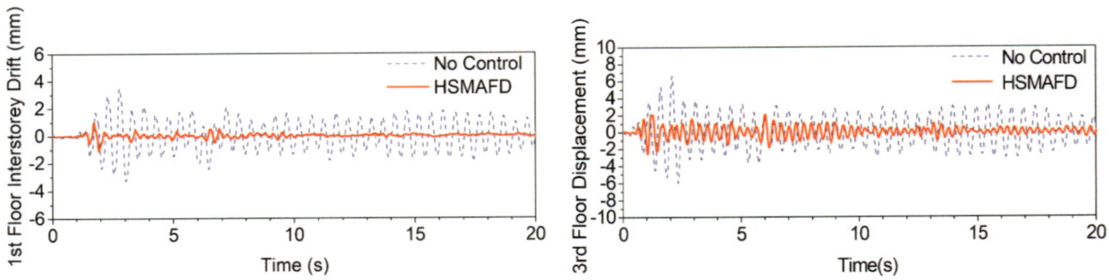

(a) El Centro record (PGA=0.15g).

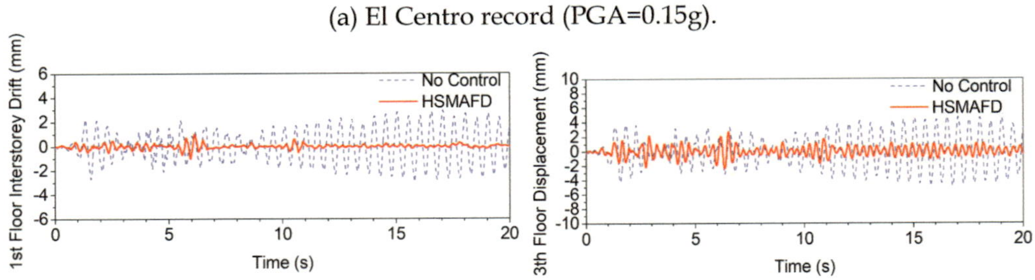

(b) Taft record (PGA=0.15g).

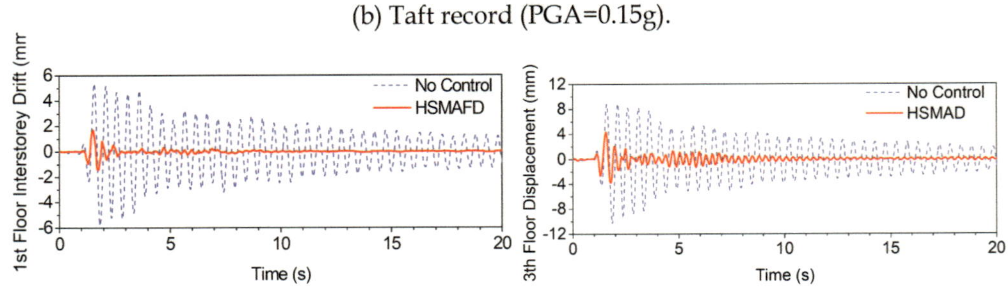

(c) Tianjin record (PGA=0.15g).

Fig. 18. Time history of interstorey drift of the 1st floor and absolute displacement of the 3rd floor under different seismic records

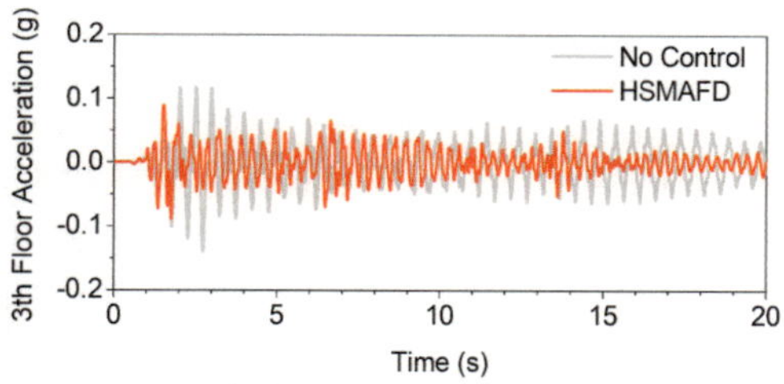

(a) El Centro, PGA=0.15g

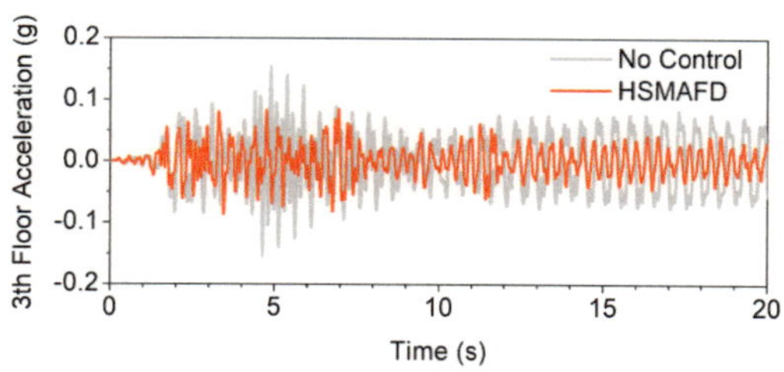

(b) Taft, PGA=0.15g

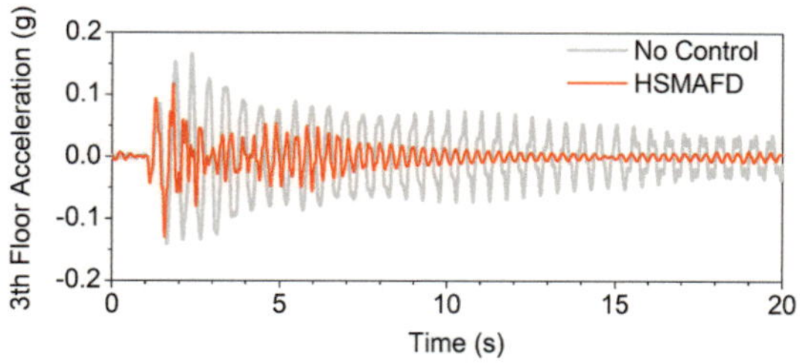

(c) Tianjin, PGA=0.15g

Fig. 19. Time history of acceleration of the 3th floor under different seismic records

Fig. 18 shows the time history of the 1st floor interstorey drifts and the 3rd floor absolute displacement of structure with and without energy dissipation system under different seismic records. In the figures, the blue dashed lines and the red solid lines represent the seismic response of the free and controlled structures, respectively. As we can seen, the seismic response of the structure have been significant reduced by the SMA damper.

Fig. 19 shows the time history of the 3rd floor acceleration of the free and controlled structure. As can be seen, the accelerations of the 3rd floor have been reduced, but no significant control effects are found because the energy dissipation system adds supplemental stiffness into the original structure.

4. SMA Damper Installation Optimization for Three-dimensional Structural Response Control with Improved Multi-objective Genetic Algorithm

4.1 Introduction

In order to use damping devices economically, the optimization of their size and position in a structure is an important issue. Researchers have suggested different practical schemes for optimal configuration of dampers, such as sequential search algorithms (Zhang et al, 1992; Shukla et al, 1999), a steepest directions search algorithm (Takewaki 1997), a worst-out-best-in and exhaustive single point substitution methods (Agrawal, et al 2000) and genetic algorithms (GAs) (Singh et al, 2002; Wongprasert et al, 2004). However, these studies are restricted to two-dimensional plane frame structures. Very few practical building structures have no eccentric axes of rotation and inertia, and can be modeled as plane structures. Moreover, a three-dimensional (3-D) structure with the eccentric location of axes of rotation and inertia has coupled lateral and torsional responses under the action of earthquake, even when excited unidirectionally. Furthermore, seismic ground motions are generally not unidirectional phenomena (Ahlawat et al,2003). Thus, these optimal schemes cannot be used for the majority of practical buildings. Some efforts to address optimal damper systems for 3-D structures have been made. Wu and Ou (1997) studied the optimal placement procedures for determining the visco-elastic dampers in asymmetrical structures to minimize the coupling effects of translation-torsion. Ahlawat and Ramaswamy (2003) designed an optimal tuned mass damper system in order to control the acceleration, displacement and torsion responses of a 3-D building. Although these designs serve well, they are carried out based on certain assumptions, i.e. the size of dampers is not included in the optimization problem but determined in advance. In reality, like the position of dampers, the size of dampers also has a significant effect on the control of structural response (Li et al, 2001). The determination of the size of dampers may be affected by many factors, for instance the total weight of structures, total cost of control, and control demands. If the size and position of dampers and the structural performance are considered simultaneously, the problem may become more complex and be characterized as a multi-objective optimization problem (MOP).

In view of mathematics, multi-objective optimization offers the potential tool to consider various and mutually conflicting requirements inherent in a problem so that it admits a family of perfect solutions rather than a single one. The GAs have successfully been applied to the MOPs. Most multi-objective GAs involve in assigning a single fitness value to each individual, weather via a weighted-sum approach that combines all objectives into a single fitness function, or some forms of ranking scheme that assigns better fitness values to individuals based on their non-dominance (Crossley et al, 1999). However, for the former, choosing the proper weights induces difficulties, because the likely output would normally be affected by the dominant objective only (Zheng et al, 2004); and for the latter, the population rank is a really complicated process for practical applications. Crossley et al (1999) presented the two-branch tournament GA that evaluates individuals competing on one of two objectives, i.e. one branch of the tournament selects individuals according to the first objective, the other branch for the second objective. This selection mechanism seems more convenient for practical use, but it fails to work well in a problem with more objectives because more branches corresponding to more objectives are required. For example, Ahlawat and Ramaswamy (2003) employed three branches instead of two to handle three objective functions, which resulted in an increasingly complicated selection and more computational effort. In addition, for a constrained optimization, penalty function methods have been successfully used in the GAs to treat single-objective constrained optimization by adding a penalty term to the fitness of an infeasible individual to make its fitness never attain that of a feasible one, but the approaches cannot be extended to a MOP, because the search goal of the MOP is a region rather than an point, and the fitness of a point has no direct connection with its objective functions(Cheng and Li, 1997).

In this section, the problem of optimizing a SMA damper system when incorporated into a 3-D structure is given. To solve a MOP, a constrained multi-objective GA is first presented by integrating an improved selection operator based on the two/three-branch tournament GA and a Pareto-based penalty function method. The improved algorithm is then followed by verification through two mathematical benchmark problems. Finally, the improved GA is used to deal with the optimal size and position of SMA dampers for a six-story torsionally coupled building subjected to earthquake in order to achieve a desired level of reduction in both translational and torsional responses at minimum cost.

4.2 Improved Constrained Multi-objective GA

4.2.1 Selection operator

The improved approach is constructed by evaluating individuals based on their non-dominance in one run rather than reproducing individuals on each objective in each branch as the two/three-branch tournament GA. To do this, a pot and parent pools are set in advance. The entire population of a current generation is placed in the pot. Two individuals, X_1 and X_2, are randomly selected without replacement from the pot and compete on their dominance: If $\phi_i(X_1) > \phi_i(X_2)$ for all $i \in \{1, 2, \cdots, n\}$, the individual X_2 dominates the individual X_1, X_2 is copied to the parent pool twice and X_1 is discarded, or vice versa; otherwise, the two individuals are equally dominant and each one is copied to the parent pool once. This continues until the pot is empty and the parent pool is full. Then, the individuals in the parent pool are again randomly selected without replacement, two at a time, and mated in the crossover operation to form two individuals. The mutation operation is subsequently performed on certain individuals, with a certain probability of mutation. These offspring eventually become the next generation of designs.

In the algorithm just described, there are certain guarantees about the selection process even though individuals are randomly chosen from the pot. In each run, individuals are selected without replacement, so that all individuals will have one opportunity to compete on their non-dominance, the better performing individual will always prevail against its competition, and the most copies an individual can have in the parent pool is two. These features provide a slight selection pressure toward better individuals, but still maintain the diversity needed to represent a spectrum of individuals across the Pareto optimal set (Crossley et al 1999). An example of bi-objective optimization problem is shown in Fig. 20. Four design variables are defined as X_1, X_2, X_3 and X_4 with the corresponding objective values included in the brackets, and the underlined variables represent the selected individuals. In the improved selection, all non-dominated individuals, X_1, X_2 and X_3 are always reproduced once or twice. However, the two-branch GA sometimes sees that the non-dominated individual X_1 is discarded and the dominated individual X_4 is selected twice. This example shows that the improved selection operator may be superior to the two-branch tournament in searching the Pareto optima.

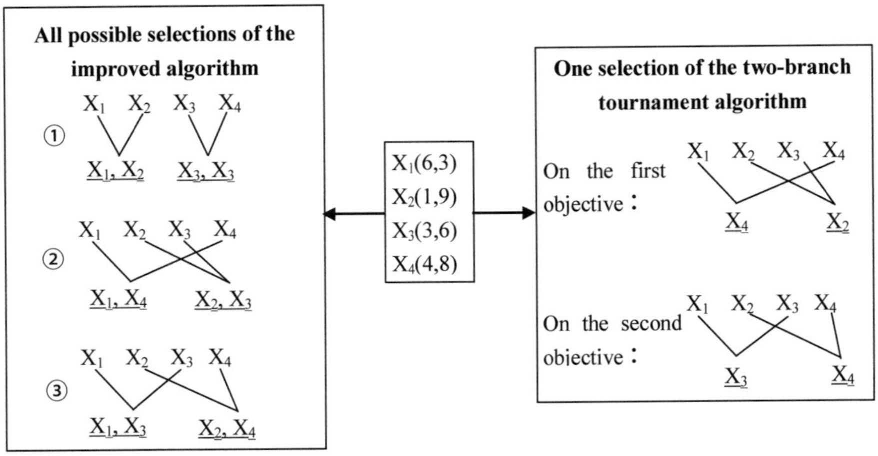

Fig. 20. Selected solutions of two-branch Tournament algorithm and its improved version

In addition, the improved selection operator takes less computational effort due to the evaluation of individuals in each generation being completed in one run, regardless of the objective dimension.

4.2.2 Pareto-based penalty function

The GAs can be applied to a constrained optimization problem by using penalty function methods to transform a constrained problem into an unconstrained one. In the selection operator above described, an individual is evaluated according to its non-dominance. Thus, the penalty functions should reflect an individual's non-dominance in a population correctly. According to the definition of the Pareto optima, the non-dominance of an individual is defined by its status (feasible or infeasible), distance from the Pareto optimal set, and position in an infeasible zone (Cheng and Li 1997). Thereby, four rules should be followed as: (1) a feasible individual is superior to an infeasible one; (2) the closer an individual is to the feasible zone, the more superior it is; and (3) the closer an individual is to the Pareto optimal set, the more superior it is; (4) the second rule supersedes the third because an infeasible solution is unacceptable in a constrained optimization problem. The Pareto-based penalty function method yields:

$$\phi_i'(X)=\begin{cases}\phi_i(X),\ r_c(X)=0, & \text{for feasible individuals}\\ C+r_c(X)+\dfrac{\phi_i(X)}{C},\ r_c(X)>0, & \text{for infeasible individuals}\end{cases},\left(i\in\{1,2,\cdots n\}\right), \quad (2)$$

where, C means a positive constant with very large value and r_c denotes the penalty term which ranks the infeasible individual according to its violated amount for constraint R_c, where

$$R_c(X)=\sum_{j=1}^{p}g_j^+(X),\ g_j^+(X)=\begin{cases}0, & g_j(X)\le 0\\ g_j(X), & g_j(X)>0\end{cases}. \quad (3)$$

If R_c equals 0, then r_c equals 0; while R_c increases, r_c increases as $1,2,\cdots$. $\phi_i(X)/C$ is the sub-penalty term that has a value within a very small range, for example $[-0.1,0.1]$.

Eq. (2) and Eq. (3) are consistent with the foregoing four rules. First, the improved selection operator guarantees that the third rule does. The new objective functions $\{\phi_1',\phi_2',\cdots,\phi_n'\}$ can indicate whether an individual is feasible ($r_c=0$) or infeasible ($r_c>0$), and can also ensure that a feasible individual gets lower objective values than an infeasible one due to the introduction of C. For an infeasible individual, the same value of r_c is used for all objective functions, which can guarantee the same penalty to all objectives. For example, design variables, X_1 and X_2, are two infeasible individuals: if X_1 is closer to the feasible zone than X_2, $R_c(X_1)<R_c(X_2)$, then $r_c(X_1)<r_c(X_2)$, so $\phi_i(X_1)<\phi_i(X_2)$ for all $i\in\{1,2,\cdots,n\}$, so X_1 dominate X_2. The sub-penalty term plays its role only when infeasible individuals are equidistant from the feasible region, due to its very small value. Seeing the just above example, if X_1 and X_2 are equidistant from the feasible region, $r_c(X_1)=r_c(X_2)$, then if $\phi_i(X_1)/C<\phi_i(X_2)/C$ for all $i\in\{1,2,\cdots,n\}$, X_1 dominates X_2, else X_1 and X_2 are equally dominant.

4.2.3 Pareto-set filter

Reproduction of a new generation cannot assure that the best traits of the parents are always inherited by their offspring. Some of these traits may never appear in future phases of evolution due to the limited population size (Cheng and Li 1997). To avoid missing optimal individuals, a Pareto-set filter is employed. At each generation, the non-dominated individuals are put into a filter. When new individuals are added, all individuals in the filter are subjected to a non-dominated check (filtering process) and the dominated individuals are discarded (Cheng and Li 1997). Thus, only non-dominated individuals are stored in the filter.

The filter size can be set to equal population size or any other reasonable size. When the desired storage capacity is exceeded, the individuals at a minimun distance relative to other individuals are removed to maintain an even design distribution in the filter (Cheng and Li 1997). The distance between individuals, X_1 and X_2, can be calculated as follows (Huang et al., 2007)

$$d=\sqrt{\sum_{i=1}^{n}\left(\frac{\phi_i(X_1)-\phi_i(X_2)}{\phi_{i,\max}-\phi_{i,\min}}\right)^2} \tag{4}$$

where $\phi_{i,\max}$ and $\phi_{i,\min}$ are the maximum and the minimum values of the ith objective function, respectively. Eq. (4) escapes potential problems when the magnitudes of the objective functions are quite different.

4.2.4 Stopping criterion

When the maximum number of generations is exceeded or the stall generation is reached, the multi-objective GA is terminated. The stall generation means that the algorithm deals with the specified number of generations with no improvement in the Pareto-set filter.

4.2.5 Flowchart of improved multi-objective GA

Fig. 21 shows the flowchart of the improved GA. In Fig. 21, "*ng*" represents the number of generations.

4.3 Optimal model of damper system for 3-D building structure

4.3.1 Equation of motion for 3-D damped building

An *n*-story 3-D shear-type building model is shown in Fig. 6. The dimensions of the building along the *x*- and *y*-direction are l and w, respectively. The center of resistance (CR) of the building does not coincide with the center of mass (CM), but for convenience, it is assumed that each center lies on a single vertical axis, with the eccentricities, e_x and e_y, in the *x*- and *y*-direction respectively. r_x and r_y are the distances from the CM to the frame columns 4 and 1, respectively. The mass and moment of inertia of each storey, m_i and ι_i, are assumed to be concentrated at each floor level, thus the building model can be idealized to have three degrees of freedom at each floor, i.e. two orthogonal lateral drifts with respect to the ground, u_{xi} and u_{yi}, and a rotation with respect to a vertical axis. $u_{\theta i}$. k_{xi}, k_{yi} and $k_{\theta i}$ are the resultant stiffness of the building in *x*-, *y*-, and θ-direction, respectively. Ground accelerations applied in *x*- and *y*-direction are denoted by the terms $\ddot{u}_{xg}$ and $\ddot{u}_{yg}$, while the torsional component is regarded as zero. Dampers may be arranged on the outer frames and provide the control forces f_{1i}, f_{2i}, f_{3i} and f_{4i} on the corresponding frames. In the definitions of the above variables, the subscript *i* denotes the *i*th floor.

The equation of motion for this idealized torsionally coupled n-story building with added dampers can be expressed as,

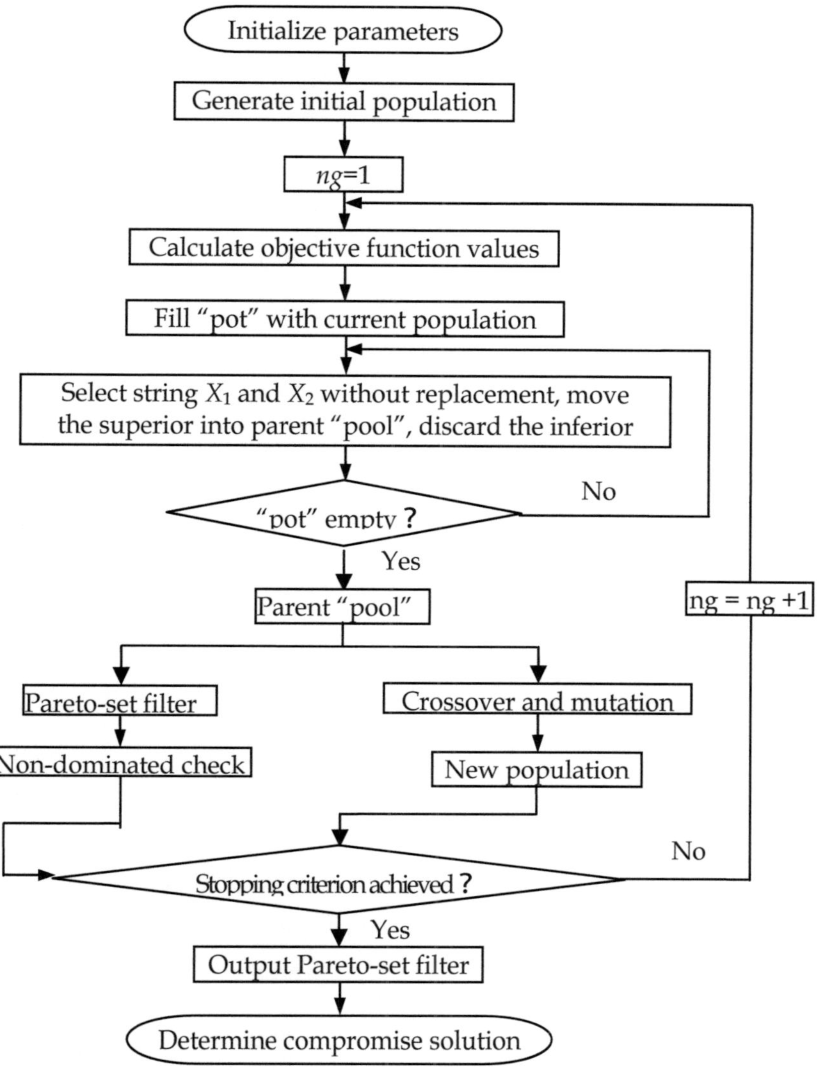

Fig. 21. Flowchart of improved constrained multi-objective GA

$$M_s\ddot{U}(t)+C_s\dot{U}(t)+K_sU(t)=-M_sD\ddot{U}_g(t)+HF(t) \tag{5}$$

where $U(t)=\{u_{x1}(t),u_{x2}(t),\cdots,u_{xn}(t),u_{y1}(t),u_{y2}(t),\cdots,u_{yn}(t),u_{\theta 1}(t),u_{\theta 2}(t),\cdots,u_{\theta n}(t)\}^T$ is the drift vector of the building with respect to the ground, $\ddot{U}_g(t)=\{\ddot{u}_{xg}(t),\ddot{u}_{yg}(t)\}^T$ denotes the earthquake induced ground acceleration vector, $D=[I\ \ \mathbf{0}\ ;\ \mathbf{0}\ \ I\ ;\ \mathbf{0}\ \ \mathbf{0}]$ represents the influencing coefficient matrix, I and $\mathbf{0}$ imply the n-column vectors with the elements 1 and zero, and M_s, K_s and C_s mean the mass, stiffness and Rayleigh damping matrices, respectively, given by,

$$M_s = diag(M_x, M_y, M_\theta), \tag{6}$$

in which $M_x = M_y = diag(m_1, m_2, \cdots, m_n), M_\theta = diag(\iota_1, \iota_2, \cdots, \iota_n)$.

$$K_s = \begin{bmatrix} K_{xx} & \tilde{\mathbf{0}} & K_{x\theta} \\ \tilde{\mathbf{0}} & K_{yy} & K_{yy}e_x \\ K_{\theta x} & K_{yy}e_x & K_{\theta\theta} \end{bmatrix}, \tag{7}$$

in which $K_{x\theta} = (K_{\theta x})^T = -K_{xx}e_y$, $K_{y\theta} = (K_{\theta y})^T = K_{yy}e_x$, and

$$K_{xx} = \begin{bmatrix} k_{x1}+k_{x2} & -k_{x2} & 0 & \cdots & 0 \\ -k_{x2} & k_{x2}+k_{x3} & -k_{x3} & \ddots & \vdots \\ 0 & \ddots & \ddots & \ddots & 0 \\ \vdots & \ddots & -k_{xn-1} & k_{xn-1}+k_{xn} & -k_{xn} \\ 0 & \cdots & 0 & -k_{xn} & k_{xn} \end{bmatrix}, \tag{8}$$

where $\tilde{\mathbf{0}}$ means the $n \times n$ zero matrix, and the elements in K_{yy} and $K_{\theta\theta}$ are similar to those in K_{xx} except that the subscripts yy or $\theta\theta$ are substitutes for xx. $F(t) = \{f_{11}(t), f_{21}(t), f_{31}(t), f_{41}(t), f_{12}(t), f_{22}(t), f_{32}(t), f_{42}(t), \cdots, f_{1n}(t), f_{2n}(t), f_{3n}(t), f_{4n}(t)\}^T$ represents the control force vector of the dampers; if no dampers are installed on the jth frame at the ith floor, $f_{ji}(t) = 0$. H is the matrix specifying the locations of the control forces; if $F(t)_i = \{f_{1i}(t), f_{2i}(t), f_{3i}(t), f_{4i}(t)\}^T$ and $F_{i+1}(t) = \{f_{1(i+1)}(t), f_{2(i+1)}(t), f_{3(i+1)}(t)n,$ $f_{4(i+1)}(t)\}^T$ are the control forces of the dampers on the ith floor and $(i+1)^{th}$ floor respectively, and their corresponding location matrices are H_i and H_{i+1}, then the resultant control force for the ith floor in x-, y-, and θ-direction is written as $H_i \times F_i(t) - H_{i+1} \times F_{i+1}(t)$, in which

$$H_i = \begin{bmatrix} -1 & 0 & -1 & 0 \\ 0 & -1 & 0 & -1 \\ r_y & l - r_x & r_y - w & -r_x \end{bmatrix}, (i = 1, 2, \cdots, n). \tag{9}$$

H is derived from integrating H_i.

Due to the translation-torsion coupling effects, the corners of the floors generally suffer maximum drifts. These drifts can be computed according to the CM's drifts and the coordinates of the corners relative to the CM.

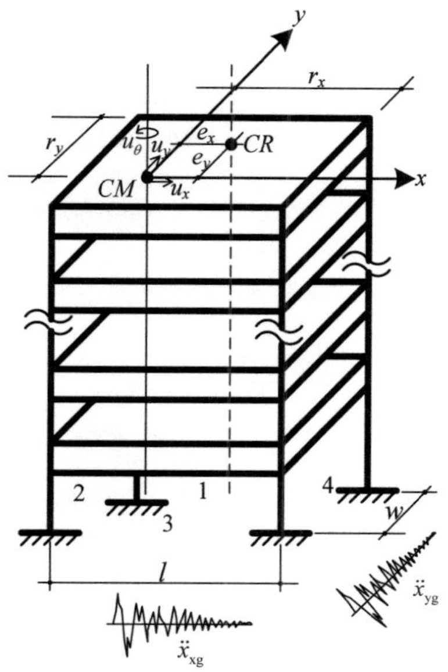

Fig. 22. Schematic representation of 3-D building

4.3.2 Optimal model of 3-D damped building

Given a damper with pre-determined mechanical properties, it is of interest to determine the size and position of such devices in a structure to maximize their effectiveness at minimum cost. This problem is presented in following multi-objective mathematical form.

In this MOP, the design variables are the size and position of dampers. Objectives are to minimize (1) the total size of the dampers to be placed in the structure, (2) the peak relative interstory drift, and (3) the peak relative torsion. These three objective functions ϕ_1, ϕ_2 and ϕ_3 are given by

$$\phi_1 = \sum_{i=1}^{n} \sum_{j=1}^{4} n_{d(ij)} \quad , \tag{10}$$

$$\phi_2 = \max_i \left(\frac{\left| u_{x(i)}^f - u_{x(i-1)}^f \right|_{\max}}{h}, \frac{\left| u_{y(i)}^f - u_{y(i-1)}^f \right|_{\max}}{h} \right) \quad , \tag{11}$$

$$\phi_3 = \max_i \left(\frac{\left|u^f_{x(i)}\right|_{\max}}{\left|u^f_{x(i)}\right|_{\text{mean}}}, \frac{\left|u^f_{y(i)}\right|_{\max}}{\left|u^f_{y(i)}\right|_{\text{mean}}}, \frac{\left|u^f_{x(i)} - u^f_{x(i-1)}\right|_{\max}}{\left|u^f_{x(i)} - u^f_{x(i-1)}\right|_{\text{mean}}}, \frac{\left|u^f_{y(i)} - u^f_{y(i-1)}\right|_{\max}}{\left|u^f_{y(i)} - u^f_{y(i-1)}\right|_{\text{mean}}} \right), \tag{12}$$

where $n_{d(ij)}$ means the size of dampers on the jth frame at the ith floor, $\left|u^f_{x(i)}\right|_{\max}$ and $\left|u^f_{y(i)}\right|_{\max}$ imply the max drift of the ith floor with respect to the ground in *x*- and *y*-direction, $\left|u^f_{x(i)}\right|_{\text{mean}}$ and $\left|u^f_{y(i)}\right|_{\text{mean}}$ denote the mean value of the maximum and minimum drift of the ith floor with respect to the ground in *x*- and *y*-direction, $\left|u^f_{x(i)} - u^f_{x(i-1)}\right|_{\max}$ and $\left|u^f_{y(i)} - u^f_{y(i-1)}\right|_{\max}$ represent the max interstory drift of the ith floor in *x*- and *y*-direction, $\left|u^f_{x(i)} - u^f_{x(i-1)}\right|_{\text{mean}}$ and $\left|u^f_{y(i)} - u^f_{y(i-1)}\right|_{\text{mean}}$ express the mean value of the maximum and minimum interstory drift of the ith floor in *x*- and *y*-direction, respectively, and h renders story height. For the three objectives, the first one is used to evaluate the reliability and economy of the damper system, and the others are adopted to quantify the safety of the building.

The constraints for the peak relative interstory drift and the peak relative torsion are formulated by,

$$g_1 = \phi_2 - [\theta] \le 0, \tag{13}$$

$$g_2 = \phi_3 - [R] \le 0, \tag{14}$$

in which $[\theta]$ is the allowable relative interstory drift , and $[R]$ means the allowable relative torsion.

4.4 Numerical Example

4.4.1 Example frame and dampers

The dimensions of the building along the x- and y-directions are 46m and 26m, respectively. Story height is 3.3m, and eccentricities $e_x = 5\text{m}$ and $e_y = 2\text{m}$. The building is assumed to be with the Rayleigh damped, with damping ratios 5% for the first two modes. Other parameters are given in Table 5.

Ren and Li (2008) developed a SMA damper with inherent self-centering behavior and enhanced energy dissipation capacity. The SMA wires have a diameter of 0.8mm with a maximum allowable strain of about 6%. One-third of the wires are pre-tensioned to a strain level of 3% while the remaining two-thirds are not pre-tensioned. The lengths of the pre-tensioned and relaxed wires are 0.385m and 0.77m, respectively, which is chosen such that the strain of the wires at a 0.007 relative interstory drift is kept below 6%. Here, the allowable value for the relative interstory drift is taken as 0.007 to ensure that no structural damage appears and the main structure functions in an approximately elastic state. In addition, given that the amount of pre-tensioned and relaxed wires in the unit damper are 4 and 2, respectively, so the size of the dampers is defined as the ratio of the practical amount of the wires to the amount of the wires in the unit damper. The mechanical parameters of the SMA wires refer to Refs. (Ren et al. 2007a; 2007b).

Table 5. Parameters for example building

Floor	Mass (kg)	Moment of inertia $(\text{kg}\cdot\text{m}^2)$	Stiffness		
			$k_x\,(\text{N/m})$	$k_y\,(\text{N/m})$	$k_\theta\,(\text{N}\cdot\text{m/rad})$
1	18375	8.337×10^{6}	3.080×10^{7}	3.080×10^{7}	1.308×10^{10}
2	16118	7.098×10^{6}	2.643×10^{7}	2.643×10^{7}	1.063×10^{10}
3	16118	7.098×10^{6}	2.643×10^{7}	2.643×10^{7}	1.063×10^{10}
4	16118	7.098×10^{6}	2.643×10^{7}	2.643×10^{7}	1.063×10^{10}
5	16118	7.098×10^{6}	2.643×10^{7}	2.643×10^{7}	1.063×10^{10}
6	16118	7.098×10^{6}	2.643×10^{7}	2.643×10^{7}	1.063×10^{10}

The Morgan Hill earthquake, which was recorded at Coyote Lake Dam in the Morgan Hill, CA, USA, during April 24, 1984, is applied to the structure. The ground peak acceleration in the x- direction is scaled to $3.0\,\text{m/s}^2$, and the peak acceleration in the y- direction is scaled to $2.55\,\text{m/s}^2$.

The responses of the bare structure are shown in Fig. 23. The maximum responses are seen on the 2nd floor, i.e. the peak relative interstory drift is 0.0067, and the peak relative torsion is about 1.51. Thus, in the MOP, the allowable relative interstory drift is considered to be 0.0067 and the relative torsion cannot exceed 1.4 according to Chinese Code for Seismic Design of Buildings (GB 50011-2001).

4.4.2 Parameters of GA

The improved GA encodes the design variables into a binary-alphabet chromosome. For the six-story asymmetrical building, the dampers can be installed at 24 different positions. In each position, the maximum size of the dampers is taken as 15, whereas the minimum is 0. Given a resolution between each value of 1 being desired, the chromosome string length is 96 bits. The population size is selected as 20. The probability of crossover and mutation are chosen as 0.7 and 0.0073, respectively. The GA is allowed to run 200 generations and the stall generation is taken to be 20.

4.4.3 Numerical results

The collected set of 80 non-dominated designs is shown in Fig. 24. It is observed that θ_{max} varies between 0.0049 and 0.0061 and R_{max} varies between 1.27 and 1.4, which indicates that the algorithm can effectively control the coupled lateral-torsional responses of the building. Moreover, the trend denotes that the performance in terms of one objective can be improved at the cost of the poorer performances of other objectives, i.e. decreasing the size of the dampers results in the increased responses of the building.

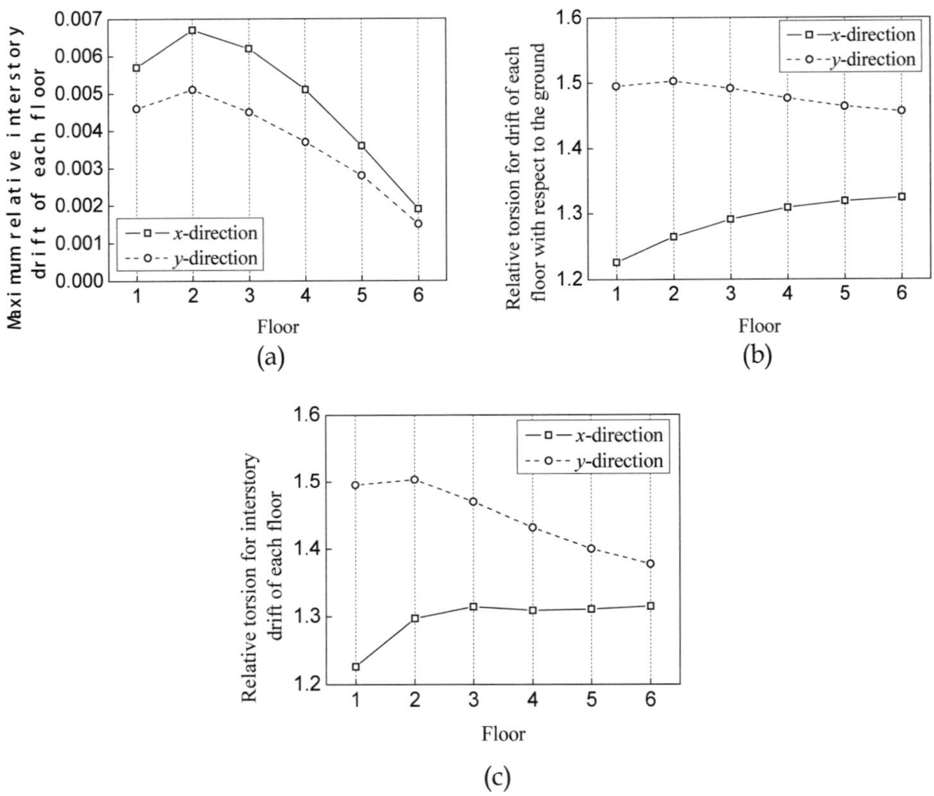

Fig. 23. Peak responses of 3-D bare structure subjected to earthquake

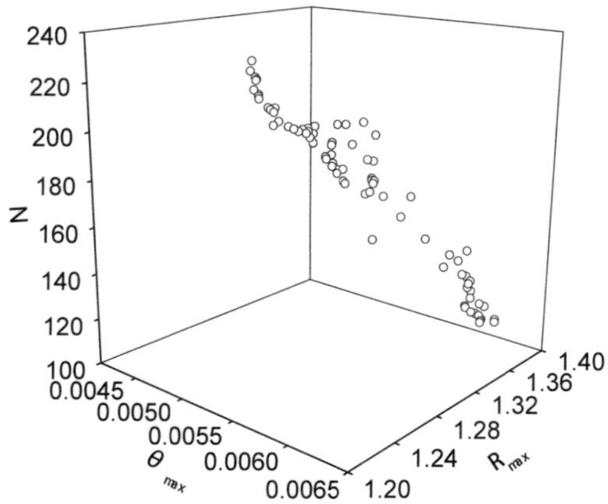

Fig. 24. Non-dominated individuals for 3-D building with SMA dampers

The non-dominated designs take information on the size and position of the dampers. Table 6 optionally lists ten optima. Taking case ① for example, the dampers with different sizes are installed on frames 1, 2, 3 and 4 at each floor: 5, 2, 12 and 2 dampers on the 1st floor, 6, 1, 9 and 1 dampers on the 2nd floor, 1, 0, 13 and 1 dampers on the 3rd floor, 3, 10, 15 and 5 dampers on the 4th floor, 1, 1, 8 and 8 dampers on the 5th floor, and 1, 4, 0 and 0 dampers on the 6th floor.

Table 6. Distribution of dampers and structural responses

Case	Distribution of dampers	$[N, \theta_{max}, R_{max}]$
①	[5 2 12 1⋮6 1 9 1⋮1 0 13 1⋮3 10 15 5⋮1 1 8 8⋮1 4 0 0]	[108,0.00609,1.3965]
②	[15 0 11 0⋮6 1 9 1⋮1 1 13 1⋮3 10 15 5⋮1 1 8 4⋮4 5 1 0]	[116, 0.00598,1.3916]
③	[6 2 12 1⋮10 1 10 1⋮1 0 14 1⋮3 10 15 1⋮2 10 15 4⋮4 5 0 9]	[137,0.00586,1.3793]
④	[14 0 15 0⋮15 8 15 0⋮1 7 13 1⋮3 10 15 2⋮1 6 8 7⋮1 4 1 0]	[147, 0.00536,1.3412]
⑤	[15 3 15 0⋮15 15 15 0⋮14 1 13 1⋮3 10 15 5⋮1 1 8 4⋮4 5 1 0]	[164,0.00516,1.3535]
⑥	[4 12 15 0⋮8 15 14 0⋮1 7 13 1⋮3 10 15 1⋮10 9 15 10⋮1 4 1 0]	[169,0.00550,1.3362]
⑦	[8 15 15 0⋮15 13 15 0⋮14 13 14 0⋮12 12 15 1⋮2 13 0 3⋮4 5 1 0]	[190,0.00499,1.3168]
⑧	[15 15 15 0⋮15 15 15 0⋮14 13 14 1⋮12 12 15 1⋮2 13 0 8⋮1 11 0 0]	[207,0.00490,1.2897]
⑨	[15 15 15 0⋮15 15 15 0⋮14 13 14 1⋮12 12 15 1⋮2 13 15 4⋮4 5 1 0]	[216,0.00487,1.2736]
⑩	[15 15 15 0⋮15 15 15 0⋮14 13 14 1⋮12 12 15 1⋮2 13 15 4⋮4 5 0 9]	[224,0.00488,1.2703]

5. Conclusions

(1) The properties of the superelastic SMA reinforced concrete members and structures are studied. The following conclusions can be drawn:

- Superelastic SMA reinforced concrete beams have good self-rehabilitation capabilities;

- Superelastic SMAs can be used as an inner energy dissipation device to change the dynamic properties of the concrete members or structures;
- Superelastic SMA reinforced structures have less horizontal and torsional responses and good self-recentering capabilities compared to reinforced concrete structures without SMAs.

(2) To assess the effectiveness of an innovative energy dissipation system (HSMAFD) in reducing the seismic response of structures, shaking table tests on a 1/4-scale, 3-story steel frame building with were carried out. The comparisons of dynamic behaviors, i.e. storey displacements, interstorey drifts and storey accelerations, of structural model with and without HSMAFD under various seismic loading were conducted.

- The fundamental frequency of the structure increases by about 50% after HSMAFD was installed due to its supplemental stiffness.
- The interstorey drifts and the absolute displacements relative to the table of the building have been non-identically reduced by HSMAFD. The first floor has a more significant effect than the other two floors because the arrangement of the HSMAFD in the first floor.
- The results show that HSMAFD were effective in suppressing the dynamic response of building structures subjected to different earthquakes by dissipating a large portion of energy through their hysteretic loops.

(3) A MOP concerning the optimal size and position of dampers in 3-D structures to minimize the translation-torsion coupling effects at minimum cost is given.

- An improved constrained multi-objective GA is developed. Based on the two-branch tournament GA, and an improved selection operator is constructed by evaluating individuals according to their non-dominance in one run, which guarantees the better performing individual winning its competition and less computational effort taken. For a constrained problem, the Pareto-based penalty function method is advanced, containing information on an individual status (feasible or infeasible), position in a search space, and distance from a Pareto optimal set, essential for success in seeking optima. The improved GA is followed by verification through two mathematical benchmark problems.
- The optimal distribution of the SMA dampers for a six-storey 3-D building subjected to earthquake is carried out. The size and position of the dampers are chosen as the design variables. The total size of dampers, the peak relative interstory drift, and the peak relative torsion are considered as the objective functions to evaluate the economy and safety of the constructed facilities. Numerical results generate a set of non-dominated solutions. Thus a suitable design can be chosen to meet the desired performance requirements.

6. References

Agrawal, A. K. & Yang, J. N. (2000). Optimal placement of passive dampers on seismic and wind-excited buildings using combinatorial optimization. *Journal of Intelligent Material Systems and Structure*, 10(12), 997-1014.

Ahlawat, A. S. & Ramaswamy, A., (2003). Multi-objective optimal absorber system for torsionally coupled seismically excited structures. *Engineering Structures*, 25(7), 941-950.

Cheng, F. Y. & Li, D. (1997). Multi-objective optimization design with pareto genetic algorithm," Journal of Structural Engineering, 123(9), 1252-1261.

Choi, E., Nam, T. H. & Cho, S. C. et al. (2008). The behavior of concrete cylinders confined by shape memory alloy wires. *Smart Materials and Structures*, 17, 1-10.

Crossley, W. A., Cook, A. M. & Fanjoy, D. W.et al. (1999). Using the two-branch tournament genetic algorithm for multi-objective design. *AIAA Journal*, 37(2), 261-267.

Cui, D., Li, H. N. & Song, G. B. (2009). Experimental investigation on dynamic properties of concrete columns embedded with shape memory alloy. *Active and Passive Smart Structures and Integrated Systems; Proc. SPIE* 7292, 729248.

Dolce, M., Cardone, D. & Marnetto, R. (2000). Implementation and testing of passive control devices based on shape memory alloys. *Earthquake Engineering and Structural Dynamics*, 29, 945-968.

Huang J Z, Wang Z & Ma R L. (2007). Multiobjective optimization design of aseismic steel frames using genetic algorithm. *Chinese Journal of Theoretical and Applied Mechanics*, 39(3), 389-397.

Li H. N., Qian H. & Song G. B. (2008). Experimental and Analytical Investigation on Innovative Compound Shape Memory Alloys Dampers for Structure Control. *Proceedings of SPIE: Smart Structures and materials & Nondestructive Evaluation and health Monitoring*. March, San Diego, California, USA.

Li, H. N., Cui, D. & Song, G. B. (2006). Experimental investigation on self-rehabilitation of intelligent concrete beams reinforced with shape memory alloys. *Proceedings of the Tenth Biennial ASCE Aerospace Division International Conference on Engineering, Construction, and Operations in Challenging Environments* (Earth and Space 2006), March, Houston, USA.

Li, Q. S., Liu, D. K. & Zhang, N. et al. (2001). Multi-level design model and genetic algorithm for structural control system optimization. *Earthquake Engineering and Structural Dynamics*, 30, 927-942.

Maji, A. K. & Negret, I. (1998). Smart prestressing with shape-memory alloy. *Journal of Engineering Mechanics*, 124 (10), 1121-1128.

Ren, W. J., Li, H. N. & Song, G. B. (2007) A one-dimensional strain-rate dependent constitutive model for superelastic shape memory alloys. *Smart Materials and Structures*, 16, 191-197.

Ren, W. J., Li, H. N. & Song, G. B. (2008) Experimental investigation and numerical evaluation of an innovative shape memory alloy damper. *Active and Passive Smart Structures and Integrated Systems; Proc. SPIE* 6928 69280W.

Ren, W. J., Li, H. N. & Song, G. B., (2007). Phenomenological modeling of the cyclic behavior of superelastic shape memory alloys. *Smart Materials and Structures*, 16, 1083-1089.

Shukla, A. K. & Datta, T. K. (1999). Optimal use of viscoelastic dampers in building frames for seismic force. *Journal of Structural Engineering*, 125(4), 401-409.

Singh, M. P. & Moreschi, L. M. (2002). Optimal placement of dampers for passive response control. *Earthquake Engineering and Structural Dynamics*, 31(4), 955-976.

Song, G., Mo, Y. L. &Otero, K. et al. (2006). Health monitoring and rehabilitation of a concrete structure using intelligent materials. *Smart Materials and Structures*, 15, 309-314.

Takewaki, I. (1997). Optimal damper placement for minimum transfer functions. *Earthquake Engineering and Structural Dynamics*, 26(11), 1113-1124.

Wongprasert, N. and Symans, M. D. (2004). Application of a genetic algorithm for optimal damper distribution within the nonlinear seismic benchmark building. *Journal of Engineering Mechanics*, 130(4), 401-406.

Wu, B., Ou, J. P. & Soong, T. T. (1997). Optimal placement of energy dissipation devices for three-dimensional structures. *Engineering Structures*, 19(2), 113-125.

Zhang, R. H. & Soong, T. T. (1992). Seismic design of viscoelastic dampers for structural applications. *Journal of Structural Engineering*, 118(5), 1375-1392.

Zheng, D. X. M., Ng, S. T. & Kumaraswamy, M. M. (2004). Applying a genetic algorithm-based multiobjective approach for time-cost optimization. *Journal of Construction Engineering and Management*, 130(2), 168-176.

9

Joining of shape memory alloys

Odd M. Akselsen
SINTEF Materials and Chemistry
Norway

1. Introduction

Shape memory alloys (SMA) have received extensive attention in the last decades due to their unique shape memory effect (Miyazaki et al, 1981), pseudo-elasticity (Miyazaki et al, 1982; Miyazaki et al, 1986), excellent corrosion properties and very good biocompatibility (Sabur et al, 1984; Takei et al, 1983; Zhi et al, 2003). So far, they have been mainly used as functional materials, with medical/biomedical applications as prime example (Morgan, 2004). When adding their ability of "learning", as well as combining the function of a sensor, actuator or that of a construction (static) material, these alloys are placed in the next higher category, the "smart" materials. Since the shape memory effect in NiTi alloys, which covers 99.99% of the market today, was discovered as late as the 1960's (Buehler & Wiley, 1965), the potential of these alloys still has not been fully understood among designers and product developers in the materials industry. Now, their use have extended to new applications in the automotive industry (e.g. diesel fuel injector), MEMS (micro-electromechanical devices), aerospace and power plants (Wu & Schetky, 2000). Moreover, SMA has been gradually introduced to the civil engineering segment, where their ability as actuators, passive energy dissipaters dampers for civil structure control have been utilized (Song et al, 2006; Janke et al, 2005). Large structures as buildings and bridges can be exposed to severe vibration, which may introduce dangerous stresses. SMA tendons may be applied to modify structures to sustain high levels of seismic vibration. By strategic placement of such tendons, structural damage may be minimized during an earthquake. This principle has also been used to preserve architectural heritage in seismic areas.

In spite of recent landmarks in SMA applications, a limiting factor in further application is their poor machinability, which often sets limits to the freedom in design. Another factor is the lack of available joining techniques, particularly for joining of SMA to other materials. Although the subject has gained more focus lately, there are still quite few publications dealing with joining as regards systematic variations in joining parameters. Moreover, the requirements set to the alloy itself should also be approached, or even better, matched after joining. The accomplishment of sound joints is very challenging, since it is not limited to the joint strength, but also comprises the shape memory effect and pseudo-elasticity.

The present report is worked out to point some focus towards challenges and recent findings in the area of joining shape memory alloys, both to themselves and to other metals.

The joining processes evaluated are arc welding, beam welding and various solid state techniques. As will be shown, all joining techniques that involve heat to the base metals tend generally to cause reduction of their initial properties, but the most severe reduction in shape and superelasticity recovery is achieved in arc welding. In joining SMA to other metals, formation of brittle intermetallic compounds is difficult to avoid.

2. Types of alloys

There are mainly three groups of SMA represented in joining research and development work:

i. Ni-Ti alloys
ii. Cu-Al alloys and
iii. Fe-Mn alloys

The NiTi alloys dominate the commercial market, because of their larger shape memory effect and better pseudoelasticity. They have superior properties with respect to ductility, fatigue, corrosion resistance, biocompatibility and recoverable strain. The Fe-Mn alloys are by far the cheapest (Janke et al, 2005), which may cause an increased market interest in them. Other alloys may also be of future interest; especially alloys with higher phase transformation temperatures will represent a supplement to the existing ones, broadening the temperature range for application of SMA. Due to the greatest attention, the current report contains most information from joining of NiTi alloys.

3. Characteristic features of SMA

When e.g. Ni-Ti is subjected to high temperature, it consists of austenite with a body centred cubic lattice, while on cooling the austenite transforms to martensite with monoclinic lattice through a displacive shear transformation. Thus, independent of alloy system, the following temperatures can be defined:

M_s = start temperature for martensite transformation on cooling
M_f = finish of martensite transformation
A_s = start temperature for austenite transformation on heating
A_f = finish of austenite transformation

The driving force for this transformation is the difference in Gibbs energy between the two phases of martensite and austenite, and depends on both the temperature and the level of externally induced stresses. These two factors are also playing the main role in the transformation mechanisms and represent the origin of the shape memory effect (SME) and superelasticity[a] (SE). The first effect, SME, is defined as a phenomenon such that an apparent plastic strain given at a temperature below A_s recovers by heating to a temperature above A_f, by virtue of the (crystallographically) reversible reverse transformation. The typical stress-strain curve of a SMA is shown in Fig. 1 as the dashed line. At the end of a

[a] Superelasticity and pseudo-elasticity are equivalent terms; both are being used.

mechanical loading-unloading path (OBC) performed at constant temperature, the material presents residual deformation (OC). However, this deformation is not caused by dislocation movement, as it is in normal metal, but by detwinning of the martensite. Therefore, the residual strain may be recovered by heating of the material (CDO), as the material transforms from martensite to austenite. SE, which is a pseudoelasticity occurring at a temperature above A_f, is caused by stress induced martensite transformation upon loading and by subsequent reverse transformation upon unloading. This crystallographic reversibility, as shown in Fig. 1 by a dashed line, is characteristic of the thermoelastic martensite transformation, which is described by a small temperature hysteresis (i.e., $\Delta T = A_f - M_s$) and mobile austenite-martensite interfaces (Saadat et al, 2002).

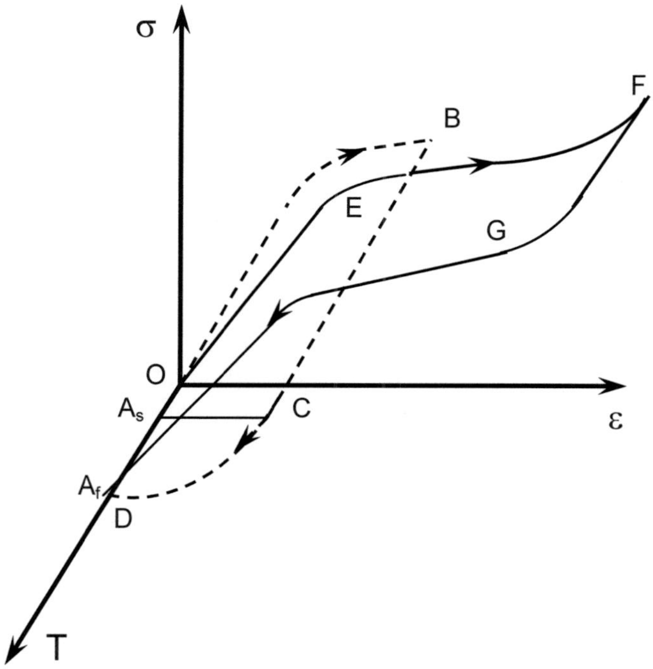

Fig. 1. Schematic stress-strain curve illustrating the shape memory effect (OBCDO) and superelasticity (DEFGD) (van der Eijk et al, 2004a).

In designing smart components in SMA, it is important to control the phase transformation temperatures in order to have the desired properties for a specific application. In Fig. 2, a schematic deformation map for a SMA showing stress-strain-phase transformation relationship. Here, M_d denotes the highest temperature below which martensite transformation can be induced by stress. Above M_d, SMA behaves like normal intermetallic compounds with CsCl-structure in which plastic deformation of the austenite occurs without inducing martensite deformation.

As indicated, the metallurgical and mechanical basics of SMA are well developed and understood. So far, however, their applicability has been limited due the lack of available fabrication techniques. Included are joining processes for assembling SMA to themselves, and not at least, to other materials. The development of such cost efficient and quality processes will probably be the key to further increase in the engineering applications of SMA. In the present report, it will be demonstrated that such joining techniques are now extensively examined, primarily for NiTi alloys, but also for other alloys such as Cu-based and more recently, Fe-based SMA. It will be shown that some processes are very promising, but also that there is more work needed to be able to maintain the initial base material properties. In fact, it may be wise to include the joining in the design for certain applications, where shape memory effect or superelasticity is of primary concern. Such conceptual design will be much more cost effective since it may be possible to integrate alloy manufacture and joining operations.

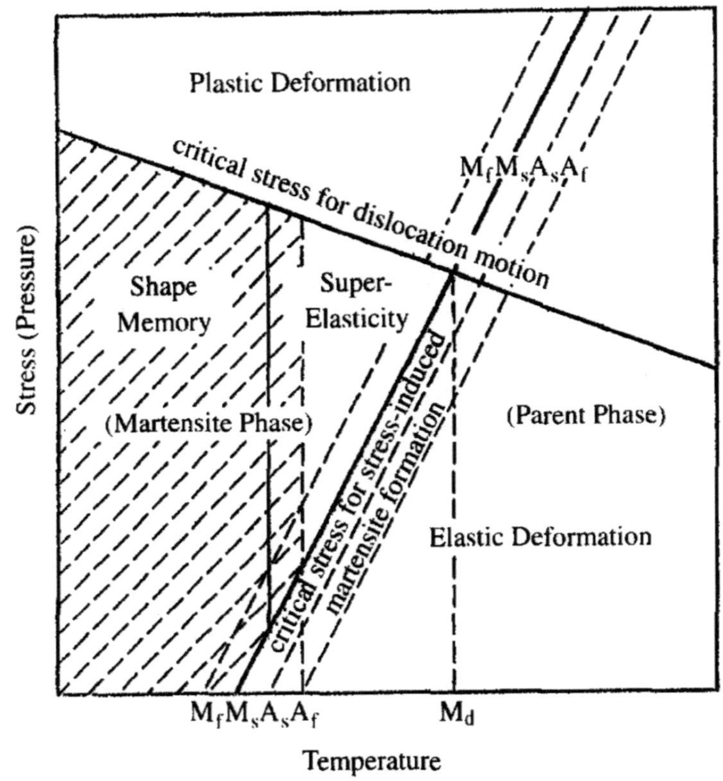

Fig. 2. Schematic deformation map illustrating phase stability, transformation temperatures and critical stress (Hosoda et al, 1998).

4. Arc welding

In the case of welding SMA, the properties to be matched between the base metal and the joint are the chemical composition and the microstructure, which dictates the shape memory response. This is where, fundamentally, the joining of SMA deviates from that of conventional metals such as steel, aluminium and titanium. In conventional alloys, successful joints are obtained in spite of differences in the chemical compositions and the microstructure of weld metal and heat affected zone (HAZ) on one hand, and the base metal on the other. In welding SMA, this may not be the case, matching of chemical composition and microstructure, as well as transformation temperatures, is very important, since this controls the mechanical properties and the behaviour of the joint.

There are many arc welding processes available, but a very limited number of them have been tested for SMA. Due to the absence of adequate welding consumables, welding has been performed by gas tungsten (GTAW) or plasma arc welding (PAW), both being carried out without welding wire. Unlike gas metal arc welding (GMAW), these do not depend on the use of melting wire since the arc is established between a non-consumable tungsten electrode and the workpiece. During arc welding of SMA, embrittlement may occur due to the reactions with oxygen, nitrogen and hydrogen at high temperatures. Proper use of shielding and backing gas may thus be vital criterion to obtain sound welds. In addition, precipitation of brittle intermetallic compounds such as $NiTi_2$ and Ni_3Ti during solidification of NiTi SMA can have adverse effects on both strength and shape memory characteristics of the material (Shinoda et al, 1991). An example is contained in Fig. 3 for autogeneous plasma arc welding of NiTi, indicating a dramatic strength reduction after welding (van der Eijk et al, 2003). Similar results have been found in GTAW. Post weld heat treatment and training of the weld area is necessary to recover the shape memory effect after welding (Ikai et al, 1999). When welding NiTi to other metals, the formation of brittle intermetallic compounds may reduce the initial properties, Fig. 4. A number of different phases are distinguishable, as shown by the SEM backscattered image of the Hastelloy side. The composition of both sides of the weld was examined by electron dispersive spectroscopy (EDS) analyses. Here, six different phases were observed, having a wide range of chemical composition.

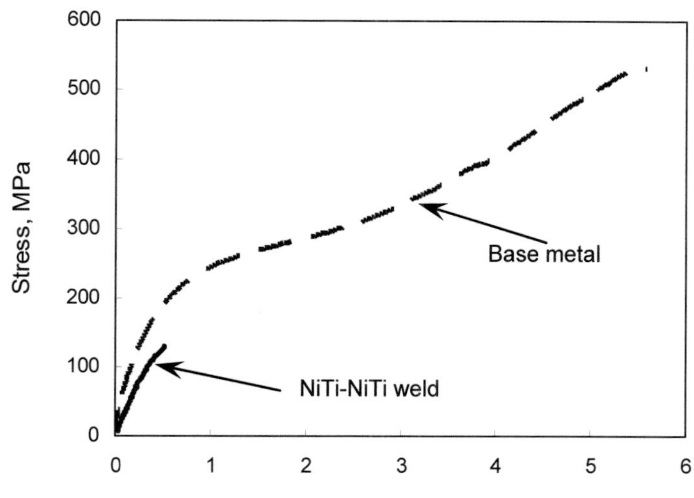

Fig. 3. Stress-strain curve of NiTi before (full line) and after GTA welding (dotted line) (van der Eijk, 2004b).

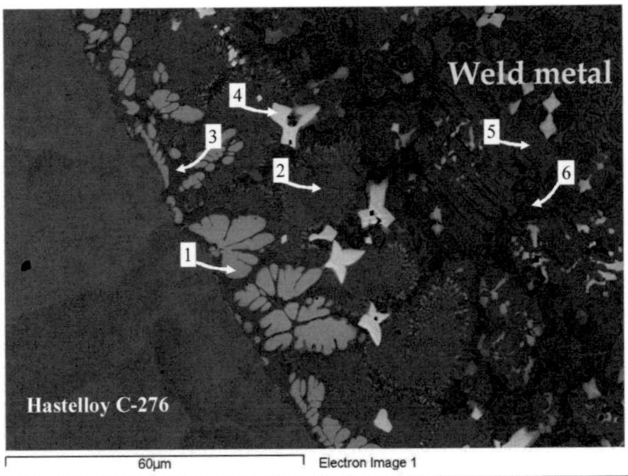

No.	Ni	Ti	Cr	Fe	Mo	W	Zr	Mn
1	30.3	4.4	20.7	4.2	30.7	9.6		
2	58.9	15.1	11.1	3.5	9.6	1.8		
3	68.4	16.3	7.3	3.1	3.8	1.6		
4	3.5	6.7	9.5		52.4	26.5	1.5	
5	75.5	21.8	1.4	1.3				
6	55.1	33.1	5.7	3.3	2.3			0.6

Fig. 4. SEM backscattered image of the fusion line of the NiTi/Hastelloy C-276 weld (van der Eijk, 2003) with corresponding EDS (wt%) analysis.

NiTi has a lower melting point (~ 1310°C) than Hastelloy C-276 (~ 1370°C) which can explain the absorption of elements from the superalloy into the NiTi. In addition, the different physical and mechanical properties may result in excessive thermal stresses and strains, and substantial cracking may occur, as shown for a plasma arc weld in Fig. 5.

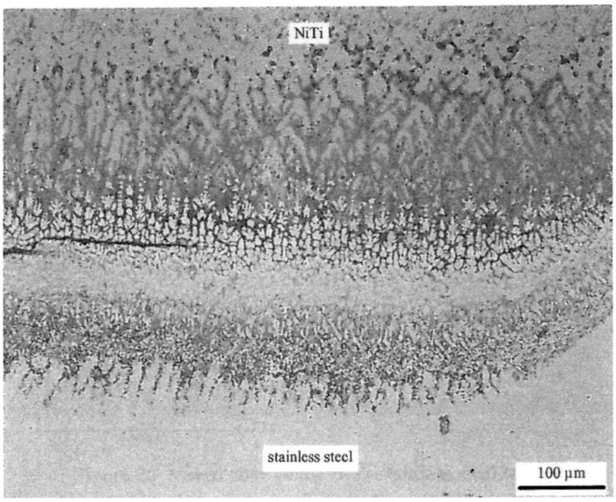

Fig. 5. Transition zone cracking in plasma arc welding of NiTi to stainless steel (van der Eijk, 2004b).

As discussed in the introduction part, the phase transformation temperatures may change in welding, which may restrict the application range of the alloy, or simply reduce its functionality. An example is contained in Table 1, showing values of M_s, M_f, A_s and A_f before and after welding. It is particularly the completion of transformation that changes due to welding, i.e., the M_f is lowered (30°C) and A_f is increased (16°C), resulting in a more sluggish phase transformation.

Welding method	Alloy	A_s (°C)	A_f (°C)	M_s (°C)	M_f (°C)
TIG (van der Eijk, 2004b)	Base metal	63	98	53	*35*
	Fused metal	59	114	65	*5*
Laser beam (Falvo et al, 2005)	Base metal	67	93	44	*10*
	Fused metal	*81*	*115*	*70*	*35*

Table 1. Results from DSC measurements of base metal and fused metal in NiTi/NiTi welds.

5. Beam welding

Laser beam welding results in narrower weld zones than arc welding (Pfeifer et al, 2008), as illustrated in Fig. 6. Both CO_2 gas laser (Tuissi et al, 2003; Hsu et al, 2001) and more extensively Nd:YAG solid state laser (Schlossmacher et al, 1994; Schlossmacher et al, 1997a; Schlossmacher et al, 1997b; Haas & Schüssler, 1999; Tuissi et al, 1999; Yan et al, 2006; Gugel et al, 2008; Khan er al, 2008; Song et al, 2008; Maletta et al, 2009) have been used. The Nd:YAG laser may give smaller welding spots and higher penetration depth than CO_2 laser since the plasma shielding effect is reduced due to the shorter wavelength (Haas et al, 1999).

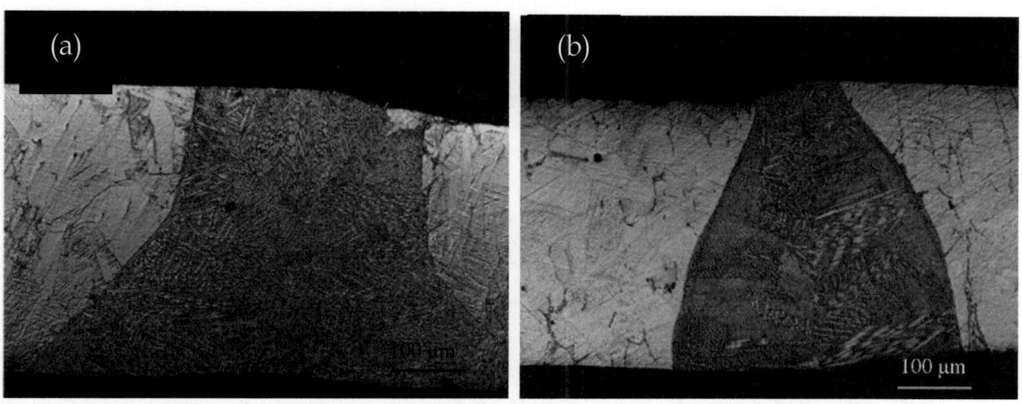

Fig. 6. Macrographs of (a) GTA weld and (b) laser beam weld; FeMnSiCr alloy (Dong et al, 2006).

As for arc welding, gas shielding is important to protect the fused metal against reaction with oxygen, nitrogen and hydrogen. Moreover, the weld quality will depend on the welding parameters (power, travel speed). The effect of laser power on minimum weld width is illustrated in Fig. 7, where the width is increased around 200 μm when the power is raised from 0.6 to 0.9 kW. The width of the fused zone, and the HAZ, is important for the thermal stresses and strains (usually called residual stresses and strains) which tend to increase with increasing width due to the larger volume to expand during heating and contract during cooling. With lower heat input, an associated finer microstructure will be expected. This should help strength enhancement through the Hall-Petch relationship. However, as for arc welding, the strength reduction may be quite severe, as illustrated in Fig. 8. The fracture took place in the weld region.

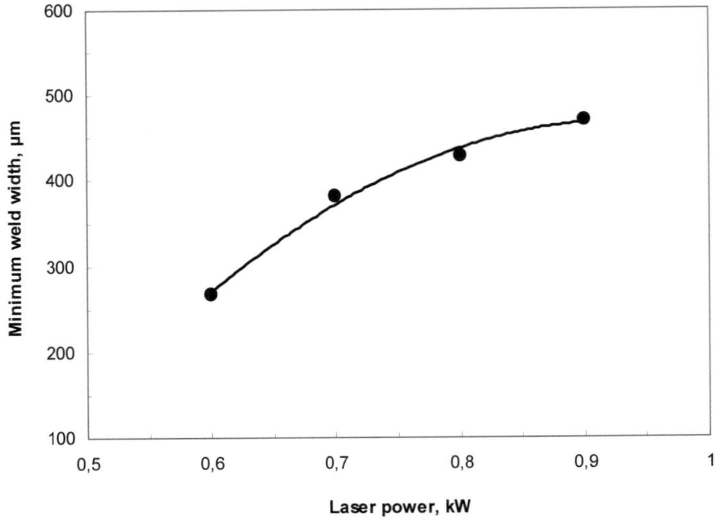

Fig. 7. Effect of laser power on minimum weld width (Khan et al, 2008).

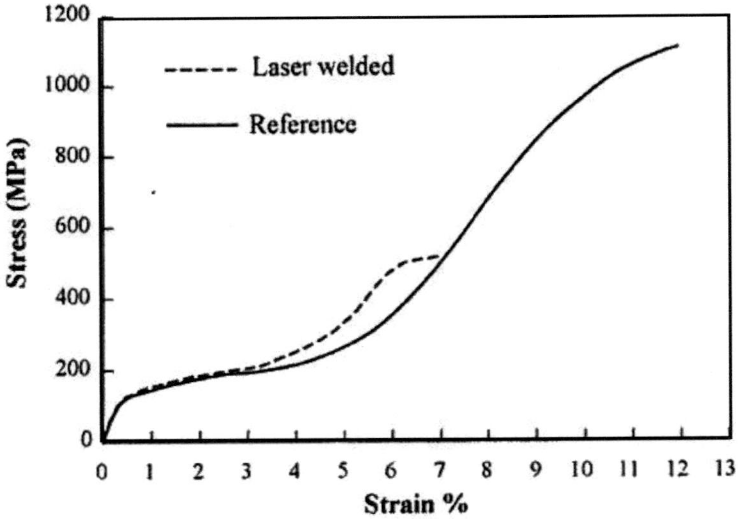

Fig. 8. Tensile stress-strain curve of NiTi before (full line) and after laser welding (dotted line). (Falvo et al, 2005).

The as welded properties can be improved by training. This point is outlined in Fig. 9, showing the measured two way shape memory strain ε_{tw}, recovery strain ε_{re} and plastic strain ε_p versus the number of training cycles before and after welding. It is seen for the base metal (filled symbols) that all strain types increase with increasing number of training cycles.

After welding, the behaviour is somewhat different. It appears that the two way shape memory strain ε_{tw}, and the recovery strain, $\varepsilon_{re,}$ pass through a maximum at 5 training cycles. Beyond 5 cycles, these strains are reduced, while the plastic strain, $\varepsilon_{re,}$ continuously increases.

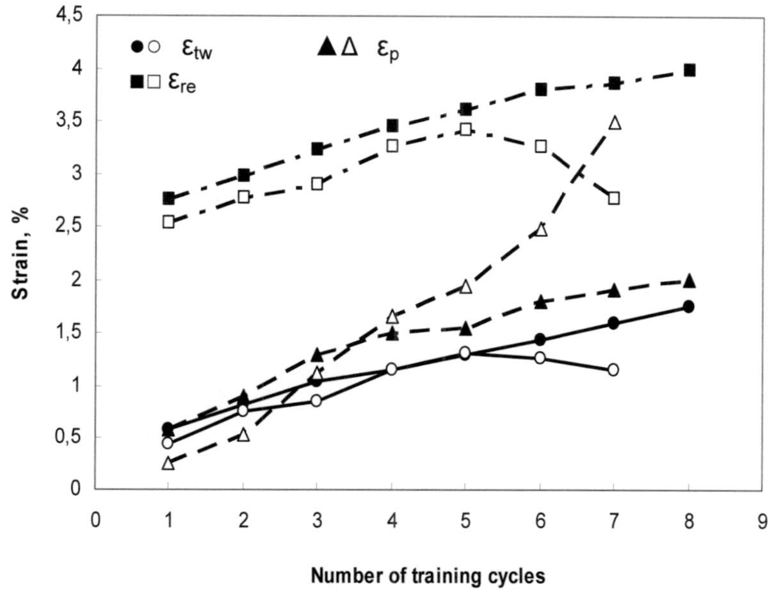

Fig. 9. Effect of training cycles on shape memory behaviour and plastic strain of NiTi alloy. Filled symbols: Base metal; Open symbols: After laser welding. (Falvo et al, 2008).

Beside the changes in mechanical properties following welding, a shift in the phase transformation temperatures (martensite to austenite and austenite to martensite) may take place. This may reduce the service temperature window for the SMA components. Phase transformation data for laser welding are listed in Table 1. As for GTA welding, phase transformation temperatures change after laser welding, the changes being even larger for laser welds. These results indicate that certain post weld heat treatment may be required to recover the initial transformation behaviour. Such information is still lacking, but it should be noticed that the post heating temperature is important.

At least to some extent, the loss of initial properties of NiTi alloys following welding, can be reduced or minimized through the so called additive laser welding (Zhao et al 2008; Zhao et al, 2010). Ce was fed by pre-placement of a foil (TiNi with 2 at% Ce) between the two thin plates of NiTi. Nb was supplied through assembling NiTi to a specimen of NiTi with 9 at% Nb. The resulting effect of these additions after laser welding is to refine the solidification structure. Without Ce or Nb additions, the solidification structure is typically columnar grains growing from the fusion line towards the weld centre line, Fig. 10a. Growth of crystals will always proceed parallel to the steepest temperature gradient in the melt; i.e., perpendicular to the isotherms. Such crystals will quickly outgrow those grains with less favourable orientation. This implies that the columnar grain morphology is dependent on the weld pool geometry, which again is controlled by the welding power, welding speed and the base metal thermal properties. In a laser weld, this may give almost a straight solidification line through the entire weld thickness. Since this area will be the last to solidify, it may contain most impurities, and hence, represent the weakest point. When adding Ce or Nb, this solidification pattern is broken, and the grains are equiaxed, Fig. 10b and c. This effect results in grain refinement since the growth of columnar grains is prevented. The consequence is an improvement of the stress-strain behaviour, as clearly evidenced in Fig. 11. Nb gave around 30% increases in strength, i.e., from 400 to 530 MPa fracture strength, and a rise in the strain at fracture beyond 40%. Ce was even more powerful and gave a strength increase twice as that found for the reference weld, and a strain at fracture of more than 200% better. For Ce, the mechanism seems to be related to formation of Ce oxides, which may act as nucleation sites in solidification, which means that the grain size is controlled by the volume fraction of the small oxide particles. For Nb, the mechanism is different. The second phase was identified as β-Nb, and solidifies at lower temperature (towards the completion of solidification). Thus, they will not act as nucleation sites, but rather prevent the growth of columnar NiTi grains.

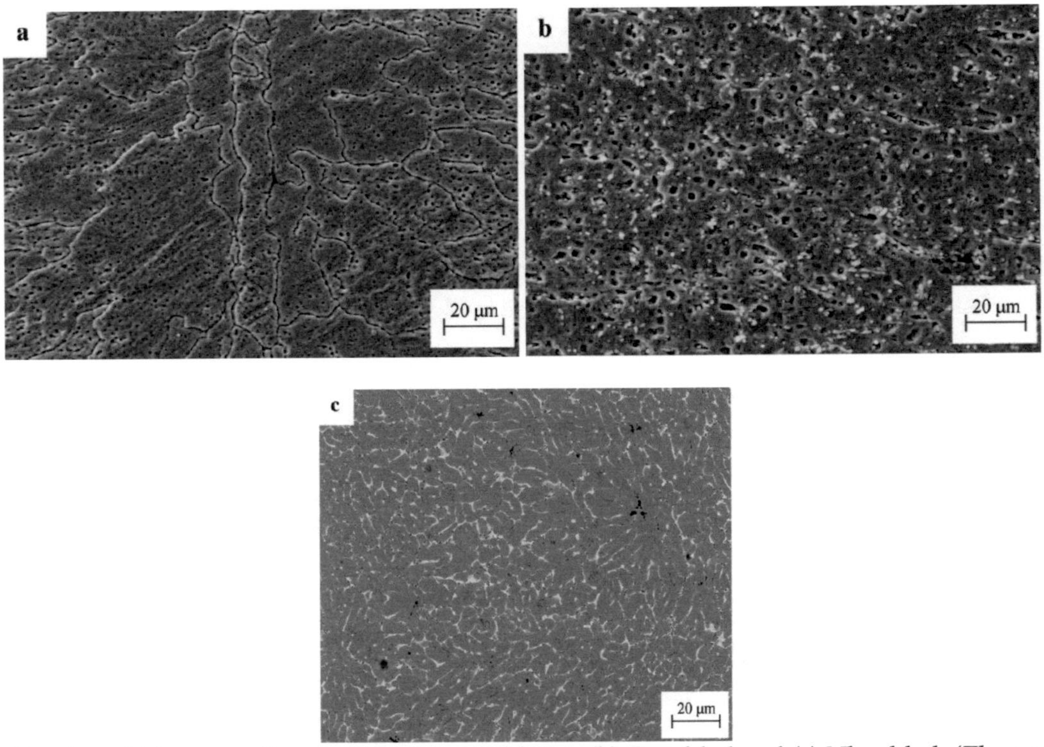

Fig. 10. SEM images of laser welds; (a) no additive, (b) Ce added and (c) Nb added. (Zhao et al, 2008).

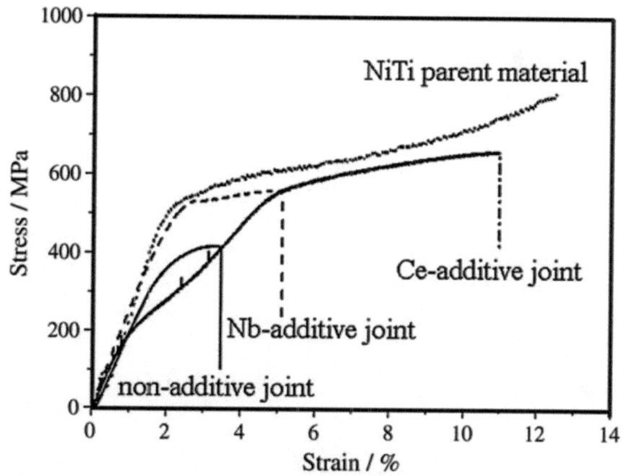

Fig. 11. Tensile stress-strain curve of NiTi before (full line) and after laser welding (dotted line). (Zhao et al, 2010).

Laser beam welding is promising in welding NiTi to stainless steel (Wang, 1997; Hall, 2003). As also observed in arc welding (van der Eijk et al, 2003 & 2004b), the stainless steel side is the one most exposed to reactions as expressed by a wider transition zone, Fig. 12. In Fig. 12a, epitaxial solidification is seen by the grains growing from the NiTi side into the weld metal.

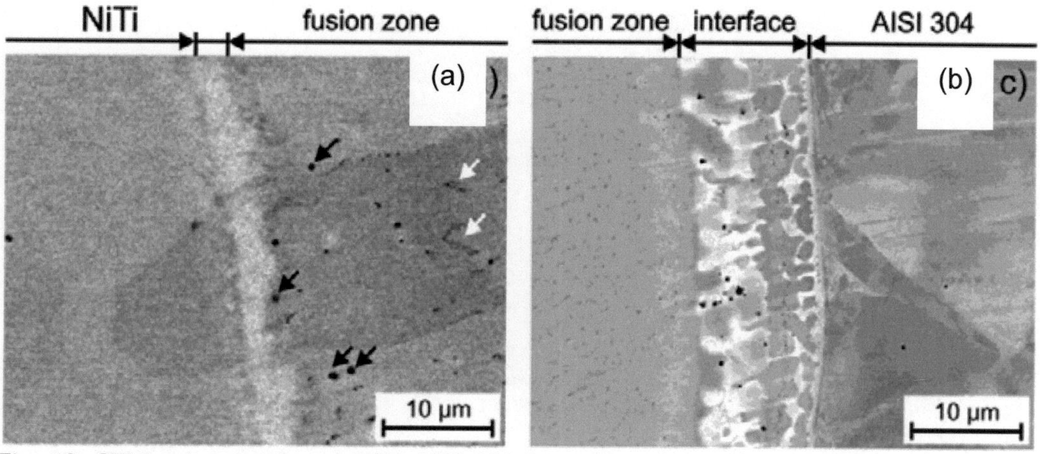

Fig. 12. SEM micrographs of NiTi-AISI 304 joint (Gugel & Theisen, 2009). (a) Interface between NiTi and fused zone, (b) interface between AISI 304 and fused zone.

The black and white arrows point out some precipitates in the weld metal. Since there is primary dendritic solidification of NiTi, these particles tend to precipitate in the interdendritic regions, being the last to solidify. An overview of the grain size distribution is illustrated in Fig. 13. The NiTi base metal consists of small grains of 30µm, while the HAZ contained smaller grains of 20 µm (Gugel & Theisen, 2009). On both sides of the weld, it is seen that the grains are nucleated at the respective fusion lines and grow inward towards the weld centre line (largest temperature gradient). A corresponding mapping of phases was also performed, and the map of the AISI 304/weld metal side is shown in Fig. 14. The dominating phase is Fe_2Ti. According to the authors, the green phase in the stainless steel side has a bcc lattice, and is probably δ-ferrite due to ferritic solidification of laser welded austenitic stainless steel. Further development of laser welding with systematic variations in the welding parameters is needed, and the effects on weld properties should be clarified.

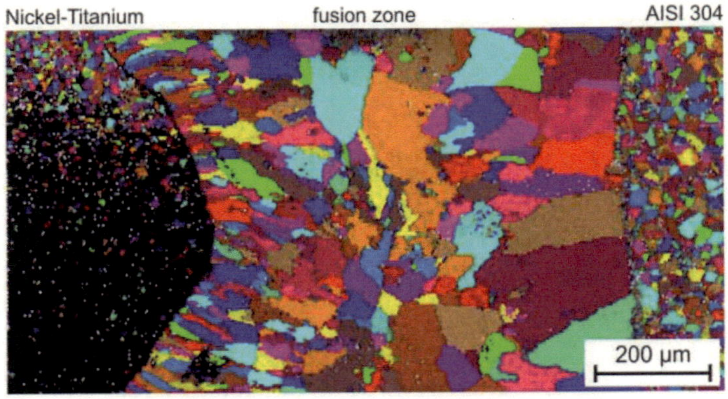

Fig. 13. Grain distribution mapping by EBSD (Gugel & Theisen, 2009).

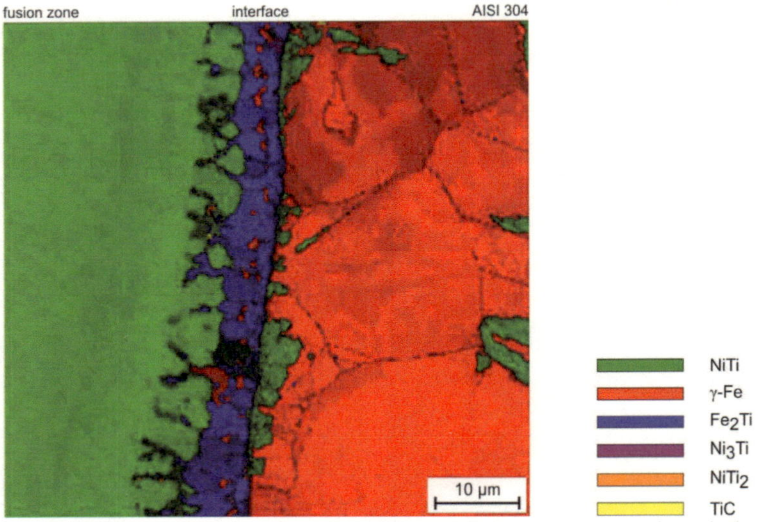

Fig. 14. Phase mapping on the interface between AISI 304 base metal and the fused zone (Gugel & Theisen, 2009).

In addition to stainless steel, laser welding (CO_2 gas laser) has been applied to joining of NiTi to the Ni-based alloy Inconel 625 and the Ni-Cu alloy Monel 400 (Tuissi et al, 2003). As found for stainless steel, extensive reactions take place at the Inconel and Monel sides of the welds. The welds gave poor strength and ductility due to transversal cracks in the as welded state. Such cracks are probably originating from welding residual stresses, in addition to the thermal expansion mismatch present. It is therefore necessary to change the chemical composition of the fused metal to reduce the cracking susceptibility. This can be done by simply adding a wire, powder or an interlayer to the fused metal during welding. Further work should be done to improve the properties of such dissimilar material joints for verification.

6. Brazing

In brazing processes, the base metals do not melt, and certain high temperature metallurgical phenomena can be avoided. These may consist of high temperature oxidation, elemental segregation and grain growth, which all may reduce the initial properties of SMA. However, even in brazing reactions will take place between the base metals and the filler alloy. Vacuum atmosphere is recommended, and the use of filler alloys containing active elements may be an advantage. An example is Ti, as indicated in Table 2 below. The table contains a survey of brazing alloys employed in this section, and AgCu alloys are the basis for all of them. In few other investigations, the use of Au and Au-based (Shiue & Wu, 2006a) or Nb (Grummon et al, 2006) filler alloys have been reported.

Alloy	Composition					Melting range, °C
	Ag	Cu	Zn	Sn	Ti	
AgCu (eutectic)	71.9	28.1	-	-	-	779
AgCuZnSn I	52	22	18	8	-	590-635
AgCuZnSn II	50-68	10-30	12-20	0-10	-	640-730
AgCuTi I	70.5	26.5	-	-	3	780-805
AgCuTi II [1]	63	35.25	-	-	1.75	780-815
AgCuTi III [2]	68.8	26.7	-	-	4.5	830-850
AgTi	96	-	-	-	4	970

[1] Cusil-ABA® [2] Ticusil®

Table 2. Brazing alloys.

As for welding, there are numerous different brazing processes, usually named after the heating method, i.e., torch, furnace, induction, dip and ultrasonic brazing, and others. For SMA in general, and NiTi particular, the alloy is quite reactive with oxygen, carbon, nitrogen and hydrogen, which set requirements to the brazing atmosphere. Thus, vacuum conditions may be preferable. To the author's knowledge, the effect of furnace pressure level on joint quality is not reported. However, brazing in 10^{-3} Pa pressure with eutectic Ag-Cu filler metal (72wt% Ag, melting temperature of 779°C) may give quite good strength and ductility, but is dependent on the overlap length. This point is shown in Fig. 15a. A considerable enhancement of strength is achieved when the overlap length is increased from 1 to 4 mm. The brazed specimen with 1 and 2 mm overlap fails at a load of 360 and 600 N, respectively, while the 4 mm overlap sample fractured at 980 N (about 820 MPa). In Fig. 15b, the recovery strain is plotted versus the total strain. Below 4% total strain (or 4% recovery strain), there is a linear 1:1 relation between the two strains, indicating 100% recovery. Beyond 4% strain, the recovery strain deviates from the 1:1 relationship with the total strain, approaching a plateau at ~ 7% for a total strain level of 10%. Obviously, a permanent loss of recovery is found.

In order to achieve a strong bond, there must be some chemical reaction following the diffusion processes between the base metals and the filler. Such reactions are the basis for brazing and cause formation of a reaction layer between the base metal and the brazing alloy. An example of resulting microstructure is shown in Fig. 16, clearly resembling the typical eutectic microstructure of Ag-Cu brazing alloys, consisting of white Ag and grey Cu rods. Note also the formation of reaction layer at the interface between the base metal and brazing alloy. The authors suggested this phase to be of the type XTi_2 (Zhao et al, 2009), where X=Ni+Ag+Cu. According to the binary phase diagram, Ni and Cu have full mutual solubility at the brazing temperature, while at low temperature (354°C), a spinodal decomposition causes phase separation into α_1 and α_2. Since there is no thermodynamic barrier to the reaction inside of the spinodal region, the decomposition is determined solely by diffusion. In the Ag-Cu and Ag-Ni binary phase diagrams, there are no intermetallic compounds formed. Moreover, there is limited solid solubility of Ni in Ag in the Ag-Ni system. By contrast, all elements (Ag, Cu, Ni) form numerous intermetallic compounds with Ti, Table 3. Therefore, the formation of intermetallics at the interface between the base metal and the brazing alloy is expected.

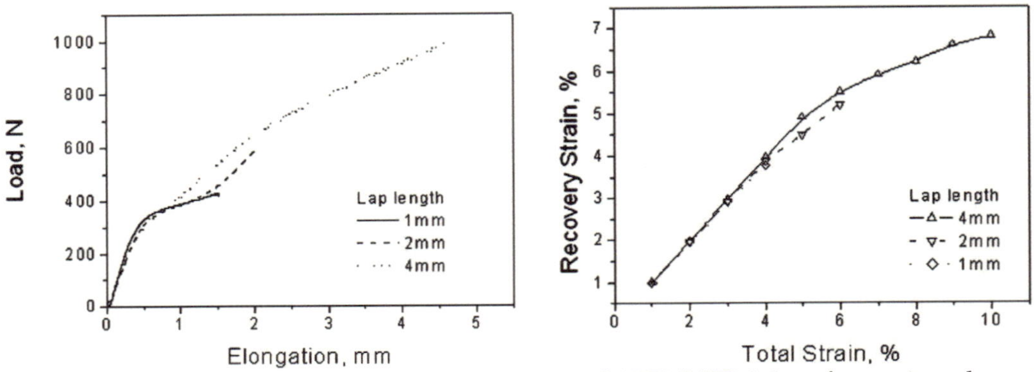

Fig. 15. Stress-strain curves and recovery strain of NiTi/NiTi joints for various brazed overlap length; brazed at 900°C for 5 min (Zhao et al, 2009).

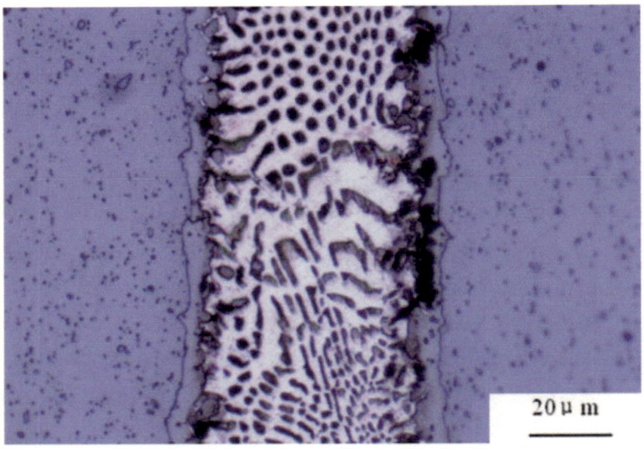

Fig. 16. Micrograph of brazed NiTi/NiTi joint (Zhao et al, 2009).

Element	Compound					
Ag	AgTi	$AgTi_2$				
Cu	Cu_4Ti	Cu_2Ti	Cu_3Ti_2	Cu_4Ti_3	CuTi	$CuTi_2$
Ni	Ni_3Ti	NiTi	$NiTi_2$			

Table 3. Intermetallic compounds in Ti_xX_y (X = Ag, Cu, Ni). (ASM Handbook, 1992)

Similar results have been found in both infrared (Shiue & Wu, 2006b) and microwave brazing (Chiu et al, 2005; van der Eijk et al, 2008). Ni may substitute Cu in XTi_2, which means that the concentration of Ni and Ti within the phase may vary quite significantly, i.e., $Ni_{67-x}Cu_xTi_{33}$. In addition, a Cu-Ti-Ni phase with some Ag was found. Finally, an Ag-rich phase will form in the brazing alloy. In summary, it is evident from experimental observations that Ag does not react with the Ni-Ti base metal. On the other hand, Cu and Ti in the brazing alloy react extensively with Ni-Ti to form XTi_2 (X = Cu+Ni, plus possibly some Ag) or Cu-Ni-Ti phases.

Using data from Villars et al (1995) and the liquidus projection of the Ag-Cu-Ti phase diagram contained in Fig. 17, the following reactions are possible in solidification of the molten Ag-Cu-Ti brazing alloy (point *A* in the figure; reference is made to the discussion by Shiue & Wu, 2006b):

At 808°C: $L + Cu_3Ti_2 \rightarrow (Ag) + Cu_4Ti$ (1)

At 783°C: $L + Cu_4Ti \rightarrow (Ag) + (Cu)$ (2)

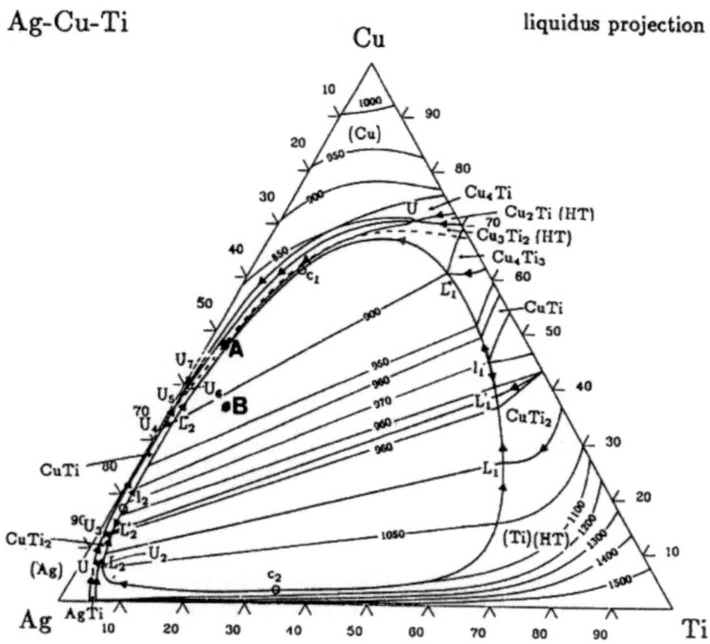

Fig. 17. Liquidus projection in ternary Ag-Cu-Ti phase diagram (Villars et al, 1995). Letters A and B refer to the brazing alloy compositions used by the authors.

On this background, the authors concluded that the microstructure of the solidified brazing alloy mainly comprised of Ag-rich and Cu-rich phases.

While NiTi has limited solubility of Ag, a certain solubility of Cu exists. Thus, both CuNiTi and (Cu,Ni)Ti may form in the Cu-Ni-Ti ternary system (Yang et al, 2004). It is also stated that, while the CuNiTi phase deteriorates the shape recovery ability, the (Cu,Ni)Ti phase preserves the shape memory behaviour (Mercier & Melton, 1979; Bricknell et al, 1979; Bricknell & Melton, 1980).

When brazing Ni-Ti to other metal, the situation is much more complex, depending on the actual metal. The Ni-based superalloy Hastelloy C-276[b] and austenitic stainless steel[c] results in extensive inter-diffusion of elements. An example is contained in Fig. 18 for NiTi/stainless steel joint. On the NiTi side of the joint there is formed a (Cu,Ni)Ti phase, while a Fe-Cr rich phase is formed on the steel side. Since reaction layer formation is a diffusion controlled process, the temperature and time used in brazing will indeed be important, in addition to the brazing alloy selection.

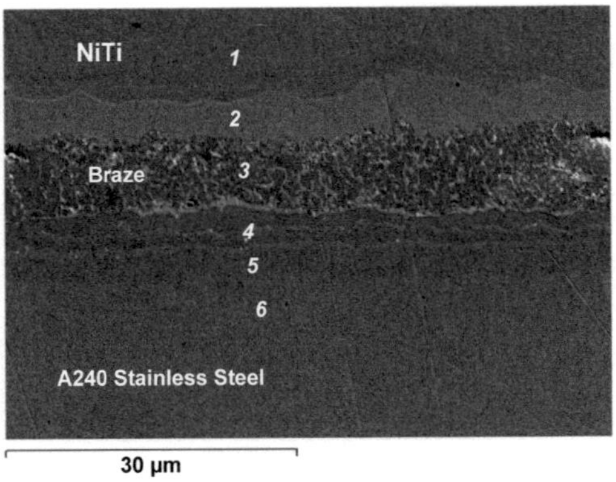

Position	Ni	Ti	Fe	Cr	Cu	Ag	Phase
1	57.3	42.7	-	-	-	-	NiTi base metal
2	28.3	26.8	2.2	-	42.7	-	(Cu,Ni)Ti
3	13.3	32.6	13.0	-	40.3	0.8	(Cu,Ni,Fe)Ti
4	6.0	3.7	59.9	30.4	-	-	FeCr-rich phase with some Ni,Ti
5	10.4	1.0	63.8	24.8	-	-	Stainless steel base metal
6	13.1	-	66.4	20.5	-	-	Stainless steel base metal

Fig. 18. SEM secondary electron image of the NiTi-stainless steel joint, brazed with Ag-Cu-Ti at 925 °C (van der Eijk et al, 2008).

In laser brazing, using the AgCuZnSn II brazing alloy in Table 2, low heat input gave low tensile strength of 190-210 MPa (Qiu et al, 2006). The strength was raised to 320-360 MPa when higher heat input was used. Keeping in mind that the initial base metal tensile strength was beyond 1100 MPa for both the NiTi alloy and the stainless steel, considerable loss in strength is caused by the brazing process. Similar findings were reported with the brazing alloy AgCuZnSn I in Table 2 (Li et al, 2006). As found in Fig. 18, interdiffusion and formation of numerous compounds take place, even more extensive than in the figure. Ag, Cu, Zn and Sn diffuse from the filler metal into both the base metals NiTi and stainless steel, while Ti and Ni from the NiTi side, and Fe, Cr and Ni from the steel side diffuse into the filler metal. The reaction layers on both sides will depend on the brazing parameters.

[b] Hastelloy C-276® contains 55wt% Ni, 14.5-16.5wr% Cr, 15-17wt% Mo, 4-7wt% Fe and 3-4.5wt% W.

[c] Containing 18-20wt% Cr, 8-10.5wt% Ni, 2wt% Mn, 1wt% Si and 0.08wt% C.

7. Adhesive bonding

Available information on adhesive bonding of NiTi is very limited. This is a bit surprising, since adhesive bonding may have limited negative influence on the initial base metal properties. Adhesives such as cyanoacrylates, epoxies, etc, can be used, when knowing the degradation susceptibility of them in different service environments. Adhesive bonding requires some surface pre-treatment to enhance the adhesion between the adhesive and the SMA base metal (Rossi et al, 2008). Such surface treatments have been applied in SMA composites consisting of NiTi wires embedded in a polymer matrix. Several surface treatment techniques have been reported, such as acid etching, polymer coating and sandblasting techniques (Paine et al, 1992) to improve the adhesion of NiTi wires to epoxy matrix. The authors concluded that sandblasting was the most efficient technique. The debonding strength was found to increase by 70%. As much as 180% improvement in debonding strength has been achieved in a similar examination (Jonnalagadda et al, 1997). Silane-coupling agents gave also 100% improvement in the adhesion strength (Smith et al, 2004). Various chemical etchants to treat NiTi fibres have been tested, but without the same strength enhancement; only 3-18% improvement was obtained (Jang & Kishi, 2005).

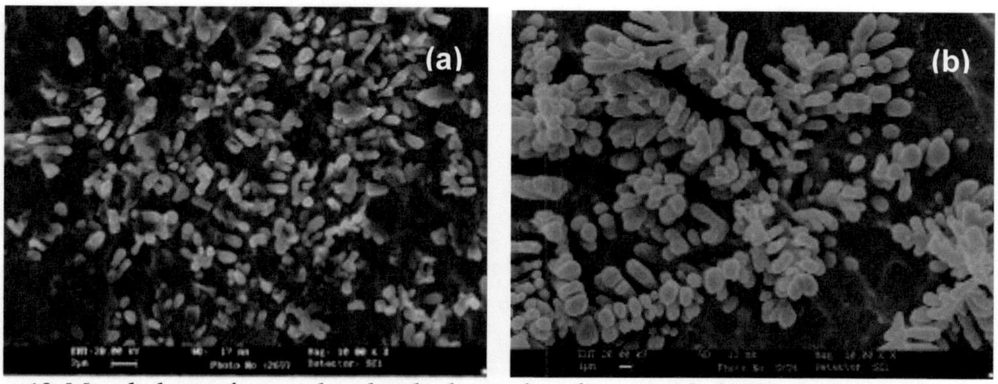

Fig. 19. Morphology of ground and etched samples of gas nitrided NiTi; (a) Etched for 5 min, and (b) etched for 30 min (Man & Zhao, 2006).

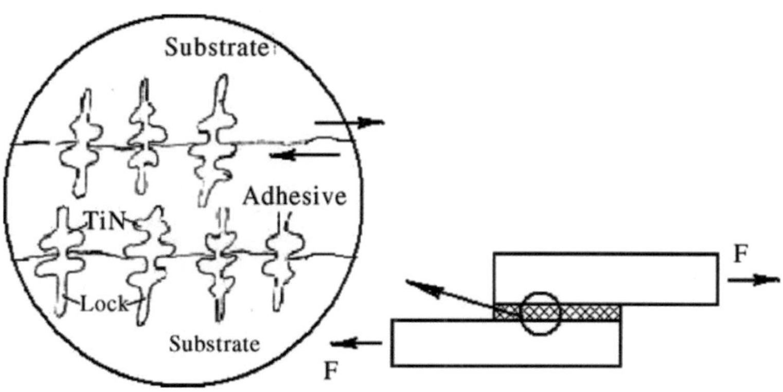

Fig. 20. Adhesive bonding (schematic) of lap joint and corresponding shear testing (Man & Zhao, 2006).

More promising then is surface coating by gas nitriding which provide hard TiN dendrites protruding from the NiTi intermetallic matrix (Cui et al, 2003; Man et al, 2005). This dendritic network, which occurs by chemical etching after the gas nitriding, gives very large increase in surface area, Fig. 19. This surface network strengthens the adhesion between the NiTi alloy and the adhesive. The concept is illustrated in Fig. 20 together with the test set-up for overlap bonding followed by shear testing. The shear test results shown in Fig. 21 reveal a considerable rise in shear strength compared with sandblasting and etching treatment. The strength level is raised from 8 to 15-20 MPa, depending on the etching time. Maximum strength of 20 MPa is found for an etching time of 30 min. These variations with etching time are closely linked to the surface morphology, which is indicated in Fig. 22.

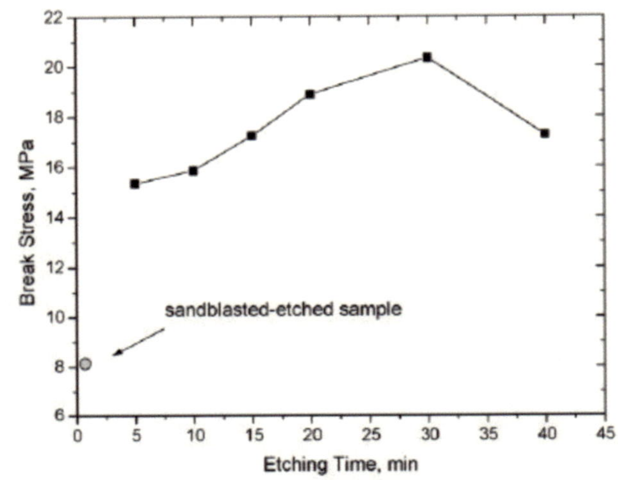

Fig. 21. Effect of etching time on lap joint shear strength (Man & Zhao, 2006).

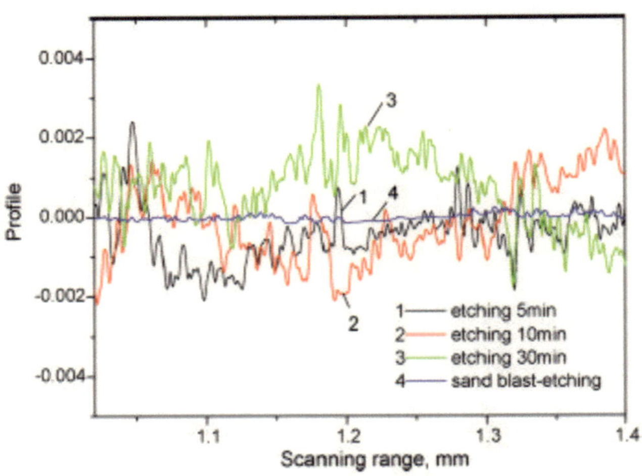

Fig. 22. Surface roughness of treated NiT (Man & Zhao, 2006).

Recent findings (Sadmezhaad et al, 2009) have also indicated that oxidation may give substantial improvement to the adhesion between NiTi and polymers, Fig. 23. Again, the adhesion strength seems to be proportional with the surface area, which is dramatically increased by the surface treatment change from type P, via types E and AO, to type O techniques.

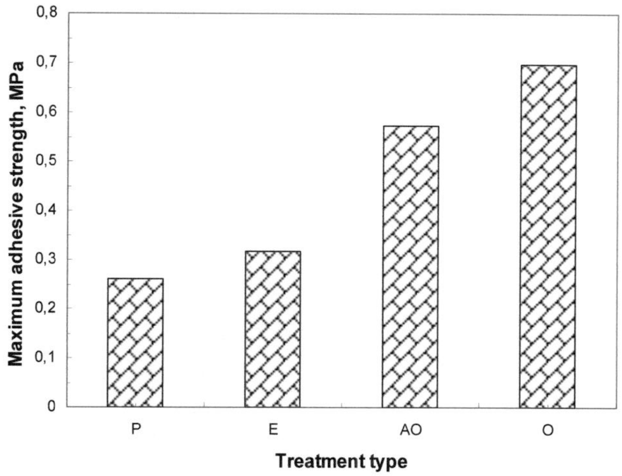

Fig. 23. Comparison of adhesive strength for different surface treatment techniques Sadmezhaad et al, 2009). (P) Straight-annealed, scraped, cleansed and washed, (E) straight-annealed, scraped, cleansed and etched, (AO) abraded and straight-annealed, and (O) straight-annealed.

8. Other techniques

Friction welding is normally an excellent process alternative when joining round bars since this joint fit-up satisfies the symmetry requirements in conventional and inertia friction welding. The first attempt to weld NiTi by friction was performed two decades ago (Shinoda et al, 1991). Although the base metal does not melt, there is substantial change in the phase transformation temperatures and loss in strength. Subsequent heat treatment at 500°C enhanced the as-welded properties to approach the same level as the base metal. This early work has not been followed up, and more information on the effects of process parameters is needed. Friction welding produces an upset collar, which needs to be removed by machining, another challenging issue.
NiTi alloys have also been friction welded to austenitic stainless steel with and without a Ni interlayer (Fukumoto et al, 2010). Without the use of Ni interlayer, a large amount of brittle Fe_2Ti intermetallic compound was formed at the weld interface. The formation of this brittle compound led to degradation of the joint strength. Use of Ni interlayer changed the microstructures at the weld interface and improved the joint strength. Fracture occurred at the interface between Ni and NiTi. The interface between Ni and NiTi was free from Fe_2Ti and consisted of mainly Ni_3Ti and NiTi.

Friction stir welding (Thomas et al, 1991) (FSW) is a special variant of friction welding used for applications where the original metal characteristics must remain unchanged as far as possible. In FSW, a cylindrical-shouldered tool, with a profiled threaded/unthreaded pin is rotated at a constant speed and fed at a constant traverse rate into the joint line between two pieces of sheet or plate material, which are butted together. Usually, the parts have to be clamped rigidly onto a backing bar in a manner that prevents the abutting joint faces from being forced apart. There are many challenges, of which the tool design and material is of great significance to the welding performance and weld quality. Tool wear may represent a huge threat against industry applications of the process. FSW has been applied to join NiTi (6.35 mm thick plates), using polycrystalline cubic boron nitride and tungsten-rhenium tool materials London et al, 2005). Similar to that measured after arc welding and beam welding, there is a change in the phase transformation temperature after welding, Fig. 24.

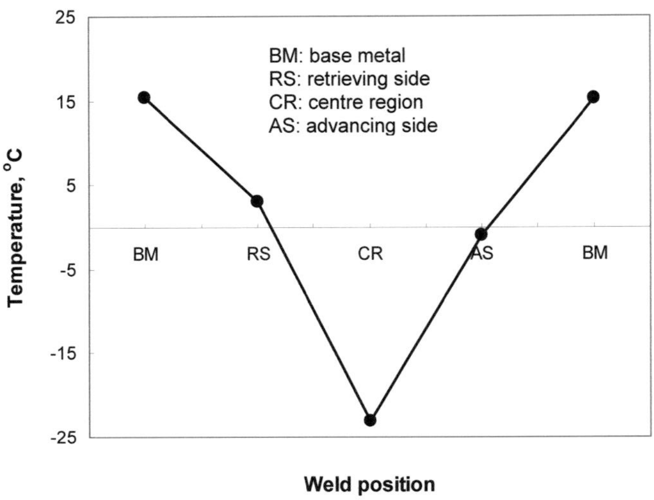

Fig. 24. Variation in austenite finish temperature across a friction stir weld (London et al, 2005).

Resistance upset butt welding has been applied for NiTi wires. Resistance welding is based upon use of electrical current and mechanical pressure to produce a weld between two parts. Welding electrodes lead the current to the two parts that are squeezed together and subsequently welded. Usually, the weld cycle must first generate adequate heat to melt a small volume, with subsequent cooling under the influence of pressure until a weld is formed with sufficient strength to keep the parts together. In conventional resistance spot and seam welding, the current density and the pressure are kept sufficiently high to form a weld, but not too high such that the material is squeezed out from the weld zone. However, in upset welding, there is an upset collar formed which usually needs to be removed by machining after welding (Nishikawa et al, 1982). The authors reported strength reduction, e.g., the tensile strength of the welds was about 80% of that of the base metal. Unfortunately, there is no follow up of this early work on resistance welding.

Explosion welding has been carried out related to manufacture of NiTi laminates. The principle is shown in Fig. 22. The joint is formed by the high velocity impact of the work pieces as a result of controlled detonation. The method consists of three main materials: (i) the base metal, (ii) the flyer, and (iii) the explosive. The main parameters in welding are the collision speed, collision angle, and the flyer speed. The interface or weld line between the base metal and the flyer is usually wavy. Apparently, this is not the case for NiTi/NiTi joint shown in Fig. 23. By this technique, NiTi can be welded to itself and to other metals. However, it is reasonable to suggest that the process will have geometric symmetry limitations, imposing restrictions to the component design.

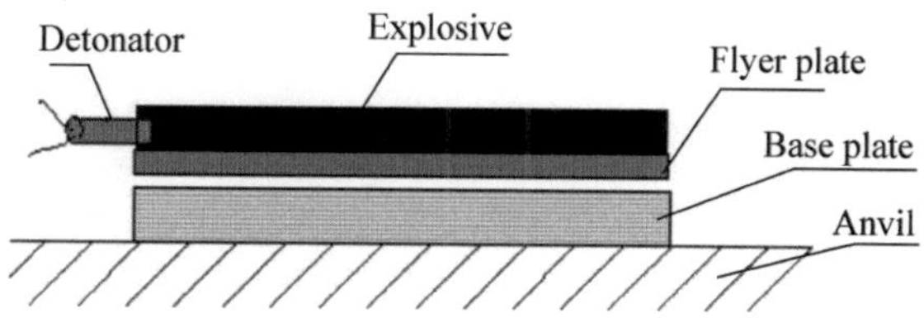

Fig. 25. Explosion welding principles.

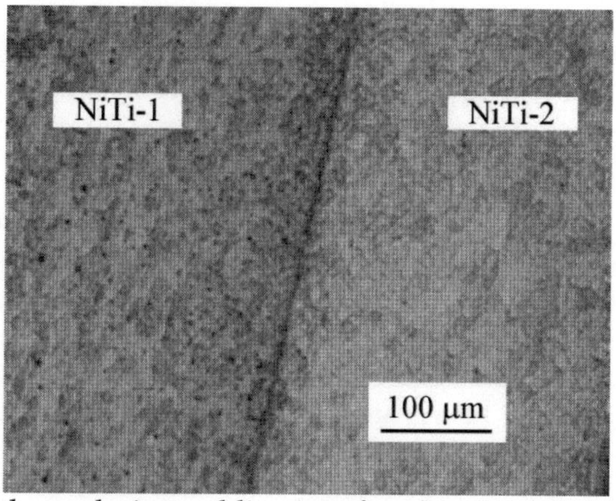

Fig. 26. Micrograph of the explosion welding interface (Yan et al, 2007).

Transient liquid phase (TLP) diffusion bonding is another possible route reported for both CuAlZn (Gomez de Salazar et al, 1997) and NiTi alloys (Kejanli et al, 2009). Diffusion bonding is a process that combines the use of temperature and pressure. The temperature range is typically 50-80% of the base metal melting temperature. The pressure is applied to provide good contact and to cause plastic deformation of surface asperities. The CuAlZn alloy was subjected to TLP diffusion bonding with an Ag interlayer. Ductile interlayers are typically used when joining dissimilar materials to accommodate strains due to different thermal expansions. The Ag interlayer was selected since Ag-Cu alloys are used as filler metals in brazing. The authors showed by their experiments that, when using their optimum bonding procedure, the shape recovery reached a value of 91% of the base metal. Also in TLP diffusion bonding of NiTi, an interlayer was used. Cu was selected, and assists in providing the transient liquid phase (low melting eutectic phase).

Soldering should also be a feasible technique for joining NiTi to itself or to other metals. An example is an early work with AgPd and AgPdGa solder alloys (Segoshi et al, 1990). Other examples are the SnAg and AuSn solders with low melting temperatures between 200 and 300°C (www.indium.com). Since SMA may be quite reluctant to wetting by solders, they may be surface treated by e.g., nickel to provide a less reactive surface (easier to wet). Alternatively, a reactive flux must be used to remove the surface oxide present. Such flux may be a non-aqueous molten salt formulated on eutectic mixtures of KOH (potassium hydroxide), NaOH (sodium hydroxide) and LiOH (lithium hydroxide), with melting temperatures in a range from about 170 to about 226° C) (Nanis, 2005). The use of an aggressive flux permits joining of SMA to other metals (e.g., stainless steel). Soldering kits are also available today, containing both the required flux and solder alloy.

It is worth to mention that shape memory alloys like NiTi is also added as reinforcements in solder alloys to improve their stress-strain behaviour. Moreover, shape memory alloys are used as couplings themselves in e.g., tube joining. Bolted connections are also made in memory metals. Such use of these alloys is outside the scope of the present report.

9. Summary and concluding remarks

In the present report, the joining of shape memory alloys to themselves or to other metals is discussed. The joining processes included are arc welding, beam welding, brazing, adhesive bonding and some other techniques like friction welding, resistance welding, explosive welding, transient liquid phase diffusion bonding and soldering.

Welding of SMA by arc welding or beam welding processes is possible but tends to produce brittle joints. Brittleness can be reduced by stress relieving, but degradation of the shape memory effect may occur due to the high temperatures that are required. Welding should also be carried under inert gases to prevent passive oxide build-up. Welding is seen to change the phase transformation temperatures, which may reduce the application window. Post welding training cycles may enhance the as welded shape memory effect and superelasticity. Laser welding tend to give narrow welds, and may thus be preferable to arc welding due to possible finer microstructure and lower thermal stresses and strains left from the welding cycle. Welding of SMA to other metals may be challenging due to the extensive formation of intermetallic compounds. These tend to be very brittle. In such cases, it may be required to use additives that prevent such formation, but very little information is available. Such information would be of great interest to expand the application of SMA-metal couplings.

Several brazing processes have been studied, including infrared furnace, laser heating and microwave heating. The main advantage of brazing is that the base metals do not melt, and that there are commercially available brazing alloys, mainly based on Ag-Cu-mixtures. Like welding, certain reduction of the initial base metal properties should be expected. Formation of intermetallic compounds at the interface between the base metal and brazing alloy will take place. This is a necessity to achieve strong bonds, but may also be negative due to the brittle nature of many of these phases. The XTi_2 (where X = Cu, Ni and/or Ag) phase is regarded to be very brittle. Brazing of SMA to other metals will give complex interdiffusion across the joint with significant reaction layer formation. The use of proper, ductile interlayer may be good choice to prevent the excessive diffusion and to accommodate thermal strains due to mismatch in thermal expansion.

Like welding, passive oxide build-up can be problematic when soldering nickel-titanium alloys. While being difficult, it can be done using SnAg or AuSn solders with an aggressive flux. Alternatively, plating with nickel or gold prior to soldering can result in good solder joints. The melting temperature of these solders is low, between 200 and 300ºC.
Transient liquid phase diffusion bonding utilizes low melting eutectics, and the process combines temperature and pressure. This eutectic is achieved by using interlayers that assist in the formation of the low melting eutectic phase. Shape recovery up to 91% has been reported.

Friction welding, both conventional and friction stir welding, has been performed. These are solid state processes, but the heavy deformation of the weld zone may cause degradation in the properties such as superelasticity, shape memory and phase transformation temperatures. Resistance welding was tested in the early stages of SMA development. The method gave promising tensile results, but has not been fully explored.

Explosion welding has been applied in NiTi-NiTi laminate production. The technique is also feasible in joining or coating NiTi to other metals, although some limitations will be present. Shape memory alloys can be joined by adhesive bonding, but surface pre-treatment is often required to enhance the bonding area. Adhesives such as cyanoacrylates, epoxies, etc, can be used, when knowing the degradation susceptibility of them in different service environments. Mechanical methods such as crimping and swaging have also been successful. The shape memory effect itself can be used and has been employed commercially in couplings.

As envisaged in the report, a great variety of joining processes has been applied to shape memory alloys. However, to some extent, there is still a lack of systematic studies on the effects of joining parameters. There is still some way to go concerning process and performance optimization, and hence, there is certainly a need for keeping the focus on joining also in the years to come.

Finally, it should be noticed that, although degradation of properties occurs in most joining processes using heat, this deterioration may take place locally. The significance of local strength or ductility changes in a component is not very clear, and there is a need for finite element modelling to simulate the impact on the component structural integrity. Moreover, with brittle materials or local brittle constituents, formation of microcracks has been observed in joining, especially when joining NiTi to other materials. The significance of such microcracks should be assessed by fracture mechanics. Some studies have been carried out (Chen at al, 2005; Daymond et al, 2007; Wang, 2007; Maletta et al, 2009), but the topic deserves more attention in future work.

10. References

ASM Handbook: Alloy Phase Diagrams, Vol.3, 1992, ASM International, Metals Park, Ohio, USA.

Bricknell, R.H., Melton, K.N. & Mercier O. (1979). The structure of NiTiCu shape memory alloys. *Metal Trans.,* 10A, 693-697.

Bricknell, R.H. & Melton, K.N. (1980). Thin foil electron microscope observations on NiTiCu shape memory alloys. *Met. Trans.*, 11A, 1541-1546.

Buehler, W.J. & Wiley, R.C. (1965). Nickel base alloys. US Patent 3174851, March 23.

Chen, J.H., Sun, W. & Wang, G.Z. (2005). Investigation on the fracture behaviour of shape memory alloy NiTi. *Metall. Mater. Trans.*, 36A, 941-955.

Chiu, K.Y., Cheng, F.T., Man, H.C. (2005). A preliminary study of cladding steel with NiTi by microwave –assisted brazing. *Mater. Sci. Eng.*, A 407, 273-281.

Cui, Z.D., Man, H.C. & Yang X.J. (2003). Characterization of the laser gas nitrided surface of NiTi SMA. *Appl. Surf. Sci.*, 208-209, 388-309.

Daymond, M.R., Young, M.L., Almer, J.D. & Dunand, D.C. (2007). Strain and texture evolution during mechanical loading of a crack tip in martensitic shape-memory NiTi. *Acta Mater.*, 55, 3929-3942.

Dong, Z.Z., Sawaguchi, T., Kajiwara, S., Kikuchi, T., Kim, S.H. & Lee, G.C. (2006). Microstructure change and shape memory characteristics in welded Fe-28Mn-6Si-5Cr-0.53Nb-0.06C alloy. *Mater. Sci. Eng.*, A 438-440, 800-803.

Eijk, C. van der, Fostervoll, H., Sallom, Z. & Akselsen, O.M. (2003). Plasma Welding of NiTi to NiTi, Stainless Steel and Hastelloy C276", *Int. Conf. Joining of Specialty Materials VI Program,* ASM Materials Solutions Conf. & Exposition, Oct. 13-15, Pittsburgh, PA, USA.

Eijk, C. van der, Zhang, Z. & Akselsen, O.M. (2004a). Seismic Dampers Based on Shape Memory Alloys: Metallurgical Background and Modelling. *Proc. 3rd European Conf. on Structural Control*, 3ECSC, July 12-15, Vienna Univ. of Technol., Vienna, Austria.

Eijk, C. van der, Fostervoll, H., Sallom, Z. & Akselsen, O.M. (2004b). Plasma Welding of NiTi to NiTi, Stainless Steel and Hastelloy C276. *Proc. Int. Conf. Advanced Metallic Materials and Their Joining,* October 25-27, Bratislava, Slovakia. International Institute of Welding.

Eijk C. van der, Sallom, Z.K. & Akselsen, O.M. (2008). Microwave brazing of NiTi shape memory alloy with Ag-Ti and Ag-Cu-Ti alloys. *Scripta Mater.*, 58, 779-781.

Falvo, A., Furgiulele, F.M. & Maletta, C. (2005). Laser welding of a NiTi alloy: Mechanical and shape memory behaviour. *Mater. Sci. Eng.*, A 412, 235-240.

Falvo, A., Furgiulele, F M. & Maletta, C. (2008). Functional behaviour of a Ni-Ti welded joint: Two-way shape memory effect. *Mater. Sci. Eng.*, A 481-482, 647-650.

Fukumoto, S., Inoue, T., Mizuno, S., Okita, K., Tomita, T. & Yamamoto, A. (2010). Friction welding of TiNi alloy to stainless steel using Ni interlayer. *Sci. Technol. Weld. Join.;* 15, 124-130.

Gomez de Salazar, J.M., Mendez., F.J., Urena, A., Guilemany, J.M. & Mellor, B.G. Transient liquid phase (TLP) diffusion bonding of a copper based shape memory alloy using silver as interlayer. *Scripta Mater.*, 37, 861-867.

Grummon, D.S., Shaw, J.A. & Foltz, J. (2006). Fabrication of cellular shape memory alloy materials by reactive eutectic brazing using niobium. *Mater. Sci. Eng.*, A 438-440, 1113-1118.

Gugel, H., Schuermann, A. & Theisen, W. Laser welding of NiTi wires. (2008). *Mater. Sci. Eng.*, A 481-482, 668-671.

Gugel, H. & Theisen, W. (2009). Microstructural investigation of laser welded dissimilar nickel-titanium-steel joints. *Proc. 8th European Symposium on Martensitic Transformations* (ESOMAT), Paper 05009, September 7-11, Prague, Czech Republic.

Haas, T. & Schüssler, A. (1999). Welding and joining of TiNi shape memory alloys: Engineering aspects and medical applications. *Proc. 1st European Conf. Shape Memory and Superelastic Technologies* (SMST '99), pp.1-11, Sept. 5-9, Antwerpen, Belgium.

Hall, P.C. (2003). Laser welding Nitinol to stainless steel. *Proc. Int. Conf. Shape Memory and Superelastic Technologies*, pp.219-228, Pacific Grove, CA, USA.

Hosoda, H., Hanada, S., Inoue, K., Fukui, T., Mishima, Y. & Suzuki, T. (1998). Martensite Transformation Temperatures and Mechanical Properties of Ternary NiTi Alloys with Offstoichiometric Compositions. *Intermetallics*, 6, 291-301.

Hsu, Y.T., Wang, Y.R., Wu, S.K. & Chen, C. (2001). Effect of CO_2 Laser Welding on the Shape-Memory and Corrosion Characteristics of TiNi Alloys. *Metall. Mater. Trans.*, 32A, 569-576.

Ikai, A., Kimura, K. & Tobushi, H. (1999). TIG Welding and Shape Memory Effect of TiNi Shape Memory Alloy. *Jour. Int. Mat. Syst. Struc.*, 7, 646-655.

Jang, B.K. & Kishi, T. (2005). Adhesive strength between TiNi fibres embedded in CFRP composites. *Mater. Lett.*, 59, 1338-1341.

Janke, L., Czaderski, C., Motavalli, M. & Ruth, J. (2005). Applications of shape memory alloys in civil engineering structures - Overview, limits and new ideas. *Mater. Struct.*, 38, 578-592.

Jonnalagadda, K., Kline, G.E. & Sottos, N.R. (1997). Local displacements and load transfer in SMA composites. *Exp. Mech.*, 37, 78-85.

Khan, M.I., Panda, S.K., Zhou, Y. (2008). Effects of welding parameters on the mechanical performance of laser welded Nitinol. *Mater. Trans.*, 49, 2702-2708.

Kejanli, H., Taskin, M., Kolukisa, S. & Topuz, P. (2009). Transient liquid phase (tlp) diffusion bonding of $Ti_{45}Ni_{49}Cu_6$ P/M components using Cu interlayer. *Int. J. Adv. Manuf. Technol.*, 44, 695-699.

Li, M.G., Sun, D.Q., Qiu, X.M., Sun, D.X. & Yin, S.Q. (2006). Effects of laser brazing parameters on microstructure and properties of TiNi shape memory alloy and stainless steel joint. *Mater. Sci. Eng.*, A 424, 17-22.

London, B., Fino, J., Pelton, A.R. & Mahoney, M. (2005). Friction stir processing of Nitinol. In *Friction Stir Welding and Processing III*, Eds. K.V. Jata, M. W. Mahoney, R.S. Mishra, T.J. Lienert, TMS, Warrendale.

Maletta, C., Falvo, A., Furgiuele, F., Barbieri, G. & Brandizzi M. (2009). Fracture behaviour of nickel-titanium laser welded joints. *J. Mater. Eng. Perform.*, 18, 569-574.

Man, H.C., Zhao, NQ.. & Cui, Z.D. (2005). Surface morphology of a laser surface nitrided and etched Ti-6Al-4V alloy. *Surf. Coat. Technol.*, 192, 341-346.

Man, H.C. & Zhao, N.Q. (2006). Enhancing the adhesive bonding strength of NiTi shape memory alloys by laser gas nitriding and selective etching. *Appl. Surf. Sci.*, 253, 1595-2000.

Mercier, O. & Melton, K.N. (1997). The substitution of Cu for Ni in NiTi shape memory alloys. *Metal Trans.*, 10A, 387-389.

Miyazaki, S., Otsuka, K. & Suzuki, Y. (1981). Transformation pseudoelasticity and deformation behaviour in a Ti-50.6 at%Ni alloy. *Scripta Metall* , 15, 287-292.

Miyazaki, S., Ohmin, Y. & Otsuka K. (1982). Characteristics of deformation and transformation pseudoelasticity in Ti-Ni alloys. *J de Phys.*, 43, 255-260.

Miyazaki, S., Igo, Y. & Otsuka, K. (1986). Effect of thermal cyclic deformation on the pseudoelasticity characteristics of Ti-Ni alloys. *Acta Metall*, 34, 2045-2051.

Morgan, N B. (2004). Medical shape memory alloy applications – the market and its product. *Mater. Sci. Eng.*, A 378, 16-23.

Nanis L. (2005). Low-temperature flux for soldering nickel-titanium alloys and other metals. US Patent 6,953,146, Oct. 11.

Nishikawa, N., Tanaka, H., Kohda, M., Nagaura, T. & Watanabe, K. (1982). Behaviour of welded part of Ti-Ni shape memory alloys. *J. de Phys.*, 43, C4, C4-839 – C4-844.

Paine, J.S.N., Jones, W.M. & Rogers, C.A. (1992). Nitinol actuator to host composite interfacial adhesion in adaptive hybrid composites. *Proc. 33rd Structural Dynamics and Materials Conf.*, AIAA, pp.556-565.

Pfeifer, R., Herzog, D., Ostendorf, A., Meier, O., Haferkamp, H., Goesling, T., Muellller, C. & Hurschler, C. (2008) Laser Welding of Shape Memory Alloys for Medical Applications. *Proc. ICALEO*, Paper #M601.

Qiu, X.M., Li, M.G., Sun, D.Q. & Liu, W.H. (2006). Study on brazing of TiNi shape memory alloy with stainless steels. *J. Mater. Proc. Technol.*, 176, 8-12.

Rossi, S., Deflorian, F., Pegoretti, A., D'Orazio, D. & Gialanella, S. (2008). Chemical and mechanical treatments to improve the surface properties of shape memory alloy wires. *Surf. Coat. Technol.*, 202, 2214-2222.

Saadat, S., Salichs, J., Noori, M., Davoodi, H., Bar-on, I., Suxuki, Y. & Msuda A. (2002). An Overview of Vibration Seismic Applications of NiTi Shape Memory Alloy. *Smart Mater. Struct.*, 11, 218-224.

Sabur, T., Yoshida, M. & Nenno S. (1984). Deformation behaviour of shape memory Ti-Ni alloy crystals. *Scripta Metall. Mater.*, 18, 363-366.

Sadrnezhaad, S.K., Nermati, N.H. & Bagheri, R. (2009). Improved adhesion of NiTi wire to silicone matrix for smart composite medical applications. *Mater. Design*, 30, 3667-3672.

Schlossmacher, P., Haas, T. & Schüssler, A. (1994). Laser Welding of Ni-Ti Shape Memory Alloys. *Proc. 1st Int. Conf. Shape Memory and Superelastic Technologies*, pp.85-90, Pacific Grove, CA, USA.

Schlossmacher, P., Haas, T. & Schüssler, A. (1997a). Laser Welding of a Ni-rich TiNi Shape Memory Alloy: Pseudoelastic properties. *Proc. 2nd Int. Conf. Shape Memory and Superelastic Technologies*, pp.137-142, Pacific Grove, CA, USA.

Schlossmacher, P., Haas, T. & Schüssler, A. (1997b). Laser Welding of a Ni-rich TiNi Shape Memory Alloy: Mechanical behaviour. *J. Phys. IV*, 7, C-251 – C-256.

Segoshi, K., Okuda, T., Okamura, K. & Yoshida, M. (1990). Soldering material for spectacle frame and spectacle frame in which said soldering material is used. US Patent 4,976,529, Dec.11.

Shinoda, T., Tsuchiya, T. & Takahashi, H. (1991). Functional Characteristics of Friction Welded Near-Equiatomic TiNi Shape Memory Alloy. *Trans. Jap. Weld. Soc.*, 22, 30-36.

Shiue, R.H. & Wu, S.K. (2006a). Infrared brazing of $Ti_{50}Ni_{50}$ shape memory alloy using gold-based braze alloys. *Gold Bull.*, 39; 200-204.

Shiue, R.H., Wu, S.K. (2006b). Infrared brazing of $Ti_{50}Ni_{50}$ shape memory alloy using two Ag-Cu-Ti active braze alloys. *Intermetallics*, 14, 630-638.

Smith, N.A., Antoun, G.G., Ellis, A.B. & Crone, W.C. (2004). Improved adhesion between nickel-titanium shape memory alloy and a polymer matrix via Silane-coupling agents. *Compos. A: Appl. Sci. Manuf.*, 35, 1312-1407.

Song, G., Ma, N. & Li, H.N. (2006). Applications of shape memory alloys in civil structures. *Eng. Struct.*, 28, 1266-1274.

Song, Y.G., Li, W.S., Li, L. & Zheng, Y.F. (2008) The influence of laser welding parameters on the microstructure and mechanical property of the as-jointed NiTi alloy wires. *Mater. Lett.*, 62, 2325-2328.

Takei, F., Miura, T. & Miyazaki S. (1983). Stress-induced martensitic transformation in a Ti-Ni single crystal. *Scripta Metall. Mater.*, 17: 987-992.

Thomas, W.M.; Nicholas, E.D.; Needham, J.C.; Murch, M.G., Temple-Smith, P. & Dawes, C.J. (1991). Friction-stir butt welding, GB Patent No. 9125978.8, International patent application No. PCT/GB92/02203.

Tuissi, A., Besseghini, S., Ranucci, T., Squatrito, F. & Pozzi, M. (1999). Effect of Nd:YAG laser welding on the functional properties of Ni-49.6at% Ti. *Mater. Sci. Eng.*, A 273-275, 813-817.

Tuissi, A., Bassani, P., Gerosa, M., Mauri, D., Pini, M., Capello, E., Previtali, B. & Vedani, M. (2003). CO_2 laser welding of NiTi/Ni-based alloys. *Proc. Int. Conf. Shape Memory and Superelastic Technologies*, 2003, Pacific Grove, CA, USA, 229-238.

Villars, P., Prince, A., Okamoto, J. (1995). *Handbook of ternary alloy phase diagrams*. Metals Park, ASM International.

Wang, G. (1997). Welding of Nitinol to Stainless Steel. *Proc. 2nd Int. Conf. Shape Memory and Superelastic Technologies*, pp.131-136, Pacific Grove, CA, USA.

Wang, G.Z. (2007). A finite element analysis of evolution of stress-strain and martensite transformation in front of a notch in shape memory alloy NiTi. *Mater. Sci. Eng.*, 460-461, 383-391.

Wu, M.H. & Schetky L.McD. (2000). Industrial applications for shape memory alloys. *Proc. Int. Conf. Shape Memory and Superelastic Technologies*, pp.171-182, Pacific Grove, CA, USA, 2000.

Yan, X.J., Yang, D.Z. & Qi, M. (2006). Rotating-bending fatigue of a laser-welded superelastic NiTi alloy wire. *Mater. Charact.*, 57, 58-63.

Yan, Z., Cui, L.S. & Zheng, Y.J. (2007). Microstructure and martensitic transformation behaviours of explosively welded NiTi/NiTi laminates. *Chinese J. Aeronautics*, 27, 168-171.

Yang, T.Y., Shiue, R.K. & Wu, S.K. (2004). Infrared brazing of $Ti_{50}Ni_{50}$ shape memory alloy using pure Cu and Ti-15Cu-15Ni foils. *Intermetallics*, 12, 1285-1292.

Zhao, X.K., Wang, W., Chen, L., Liu, F., Chen, G., Huang, J. & Zhang. H. (2008). Two-stage superelasticity of a Ce-added laser-welded TiNi alloy. *Mater. Letters.*, 62, 3539-3541.

Zhao, X.K., Tang, J.W., Lan, L., Haung, J.H., Zhang, H. & Wang Y. (2009). Vacuum brazing of NiTi alloy by AgCu eutectic filler. *Mater. Sci. Technol.*, 25, 1495-1497.

Zhao, X.K., Lan, L., Sun, H., Haung, J.H., Zhang, H. & Wang, Y. (2010). Mechanical properties of additive laser welded NiTi alloy. *Mater. Letters.*, 64, 628-631.

Zhi, C.L., Xing, K.Z. & Hong, Z. (2003). Microstructure and Superelasticity of Severely Deformed TiNi Alloy. *Mater. Letters*, 57, 1086-1090.